SpringerBriefs in Neuroscience

For further volumes:
http://www.springer.com/series/8878

Vivi M. Heine · Stephanie Dooves
Dwayne Holmes · Judith Wagner

Induced Pluripotent Stem Cells in Brain Diseases

Understanding the Methods, Epigenetic Basis, and Applications for Regenerative Medicine

Vivi M. Heine
Stephanie Dooves
Dwayne Holmes
Center for Children with White Matter
 Disorders
VU University Medical Center
De Boelelaan 1117
1081 HV Amsterdam
The Netherlands
e-mail: vm.heine@vumc.nl

Judith Wagner
Clinical Genetics
VU University Medical Center
Van der Boechorststraat 7
1081 BT Amsterdam
The Netherlands

Present address:
Judith Wagner
Adolf-Butenandt-Institute, Biochemistry
Ludwig-Maximilians-University
80336 Munich
Germany

ISSN 2191-558X
ISBN 978-94-007-2815-8
DOI 10.1007/978-94-007-2816-5
Springer Dordrecht Heidelberg London New York

e-ISSN 2191-5598
e-ISBN 978-94-007-2816-5

Library of Congress Control Number: 2011941719

© The Author(s) 2012
No part of this work may be reproduced, stored in a retrieval system, or transmitted in any form or by any means, electronic, mechanical, photocopying, microfilming, recording or otherwise, without written permission from the Publisher, with the exception of any material supplied specifically for the purpose of being entered and executed on a computer system, for exclusive use by the purchaser of the work.

Printed on acid-free paper

Springer is part of Springer Science+Business Media (www.springer.com)

Foreword

The words 'stem cells' and 'stem cell therapy' in themselves create hope: 'halting disease', 'cure of damage', 'new life'. For many years the words have been around as a promise for the future. Laymen were more confident about their applications than experts. However, the future is now approaching rapidly. Stem cell therapy is becoming a realistic target to be achieved, on a larger scale and for a growing number of applications.

The stem cell field is a fascinating, rapidly evolving field. Numerous problems and challenges are being faced and solved. The use of embryos has always been ethically controversial, but the advent of induced pluripotent stem cell technology circumvents the issue. Transdifferentiation again creates new possibilities. Step by step stem cell therapy becomes a reality.

The brain is a highly complex organ with specialized functions per area and per cell type. Advances in reprogramming and 'guided' differentiation will allow the creation of the required cell type and make cell-based therapy for the nervous system feasible.

The stem cell field is moving fast and introduces new techniques, new concepts, and new words at high speed. The present booklet gives the state of the art for 2011 with a focus on two brain disorders, for which cell-based therapy is under development. The first focus is on a neurodegenerative disorder, Parkinson's disease; the second focus is on a white matter disorder, Vanishing White Matter. The booklet is written by enthusiastic students and their supervising stem cell biologist Vivi Heine. It is a pleasure to read it. After that, the reader is up to date. For now.

Prof. Dr. Peter Heutink
Prof. Dr. Marjo S. van der Knaap

Preface

The principle goal of regenerative medicine is the restoration of damaged, dysfunctional, or missing cellular tissue, up to and including whole organs. Growing healthy replacement tissue, in vivo or in vitro, plays an important role in anticipated therapies. To generate competent replacement material, scientists confront the fundamental issues of cellular identity and plasticity.

The basis of this book is formed by the theses of three talented master students Stephanie Dooves, Dwayne Holmes and Judith Wagner. Their work discusses the recent advancements in the field of cell reprogramming. Although it is clear that we can produce pluripotent stem cells from differentiated cells, there are still a lot of unsolved issues. These issues include the efficiency and safety of reprogramming, the similarity of induced pluripotent (iPSCs) to embryonic stem cells (ESCs) and the epigenetic status of the cells. In the third chapter, the use of stem cell therapy for brain diseases will be discussed, with a focus on Parkinson's disease (PD) and Vanishing White Matter (VWM).

Dr. Vivi M. Heine

Acknowledgments

We would like to acknowledge our colleagues at the Department of Pediatrics/Child Neurology, the Medical Genomic section of the Department of Clinical Genetics, and the Center for Childhood White Matter Disorder in motivating us to write this book. In particular, we would like to thank Prof. Dr. Peter Heutink, Prof. Dr. Marjo S. van der Knaap and Dr. Gert C. Scheper for their guidance. SD and JW are grateful for their stipend of the Exchange Honours Masters Program of Neurosciences, Amsterdam-Rotterdam. JW is supported by a fellowship from the "Studienstiftung des deutschen Volkes".

Contents

1 **Reprogramming: A New Era in Regenerative Medicine** 1
   1.1 Stem Cells 1
   1.2 Reprogramming 4
        1.2.1 Viral Vectors 6
        1.2.2 Non-Viral Methods 8
   1.3 Pluripotency, Efficiency, and Identity 9
        1.3.1 Measuring Pluripotency 9
        1.3.2 Reprogramming Efficiency 10
        1.3.3 iPSCs Versus ESCs 11
        1.3.4 State of Pluripotency 13
   1.4 Differentiation 15
        1.4.1 Motor Neurons 16
        1.4.2 Dopaminergic Neurons 17
        1.4.3 Oligodendrocytes 18
        1.4.4 Astrocytes 18
        1.4.5 Selection of Differentiated Cells 19
   1.5 Trans-Differentiation 19
   References 21

2 **Understanding Epigenetic Memory is the Key to Successful Reprogramming** 27
   2.1 Epigenetic Influences on Gene Expression 27
        2.1.1 DNA Methylation 28
        2.1.2 Histone Modification 28
        2.1.3 Noncoding RNA 29
        2.1.4 Common Epigenetic Traits 30
   2.2 Epigenetic Control in De-, Re-, and Trans-Differentiation 30
   2.3 Using Epigenetics to Aid Reprogramming 32
   2.4 Epigenetics Affecting Neural Cell Fate 35
   2.5 A Caveat 37
   References 38

3 Prospects for Cell Replacement Therapies for Neurodegenerative Diseases 43
   3.1 Cell Therapies for Neurodegenerative Diseases 43
   3.2 Parkinson's Disease 44
        3.2.1 Etiology and Pathogenesis 44
        3.2.2 Treatment Options for Parkinson's Disease 45
        3.2.3 Fetal Tissue Transplantation 46
        3.2.4 Cell Therapy 47
   3.3 Childhood Brain White Matter Disorders 48
        3.3.1 Etiology and Pathogenesis 48
        3.3.2 Prospect for Cell Therapy 50
   3.4 Conclusion 51
   References 52

4 Conclusions 55
   4.1 The Promise and Limitations of iPSCs 55
   4.2 Concluding Remarks 58
   References 58

Abbreviations

5′-azac	5′-azacytidine
AA	Ascorbic acid
ACM	Astrocyte conditioned media
AID	Activation-induced cytidine deaminase
ALS	Amyotrophic lateral sclerosis
AP	Alkaline phosphatase
APC	Astrocyte precursor cell
BDNF	Brain-derived neurotrophic factor
BER	Base excision repair
BMP	Bone morphogenetic factor
CF	Cell fusion
ChIP	Chromatin immunoprecipitation
CNS	Central nervous system
CNTF	Ciliary neurotrophic factor
dbcAMP	Dibutyryl cyclic adenosine 3′,5′ monophosphate
Dnmt	DNA methyltransferase
EB	Embryoid body
ECC	Embryonal carcinoma cell
EGC	Embryonic germ cell
EGF	Endothelial growth factor
eIF	Eukaryotic initiation factor
EpiSC	Epiblast derived stem cell
ERK1/2	Extracellular signal-regulated kinase 1/2
ESC	Embryonic stem cell
FACS	Fluorescence activated cell sorting
FGF	Fibroblast growth factor
GDNF	Glial-derived neurotrophic factor
GFAP	Glial fibrillary acidic protein
GFP	Green fluorescent protein
GID	Graft induced dyskinesias
GPC	Glial precursor cell

GRM	Glial restricted media
GSK3β	Glycogen synthase kinase 3β
H..1K..2	The ..2th lysine of the histone protein ..1
H..K..ac	Acytelation of H..K..
H..K..me3	Trimethylation of H..K..
haGSC	Human germline stem cell
HAT	Histone acetyl-transferase
HCP	High density CpG islands
HDAC	Histone deacetylase
HDM	Histone demethylation
HDUB	Histone de-ubiquitinase
hESC	Human embryonic stem cell
hiPSC	Human induced pluripotent stem cell
HMT	Histone methyl-transferase
HOTAIR	HOX antisense intergenic RNA
HUT	Histone ubiquitin-transferase
ICM	Inner cell mass
IGF-1	Insulin-like growth factor 1
iPSC	Induced pluripotent stem cell
Klf4	Krüppel-like factor 4
L-dopa	Levodopa
LCP	Low density CpG islands
LIF	Leukemia inhibitory factor
lncRNA	Long non-coding RNA
maGSC	Mouse germline stem cell
MBD	Methyl-CpG-binding domain protein
MEF	Mouse embryonic fibroblast
mESC	Mouse embryonic stem cell
miPSC	Mouse induced pluripotent stem cell
miRNA	Micro RNA
MPP	Multipotent progenitors
MPTP	1-methyl-4-phenyl-1,2,3,6-tetrahydropyridine
MRI	Magnetic resonance imaging
mRNA	Messenger RNA
ncRNA	Non-coding RNA
NF1a	Nuclear factor 1a
NPC	Neural progenitor cell
NRSF	Neural-restrictive silencer factor
NSC	Neural stem cell
Oct4	Octamer binding transcription factor 4
OPC	Oligodendrocyte precursor cell
OSK	Oct4, Sox2, Klf4
OSKM	Oct4, Sox2, Klf4, c-Myc
PC	Progenitor cell
PcG	Polycomb group

PRC	Polycomb repressive complex
PD	Parkinson's disease
PDGF	Platelet-derived growth factor
PET	Positron emission tomography
PGC	Primordial germ cell
pre-miRNA	Precursor microRNA
primiRNA	Primary microRNA
RA	Retinoic acid
RISC	RNA-induced silencing complex
RT-PCR	Reverse transcription polymerase chain reaction
SAHA	Suberoylanilide hydroxamic acid
SC	Stem cell
SCNT	Somatic cell nuclear transfer
SHH	Sonic hedgehog
siRNA	Small interfering RNA
SNM	Spherical neural masses
Sox2	Sex determining region Y-box 2
STAT3	Signal transducer and activator of transcription 3
TH	Tyrosine hydroxylase
TF	Transcription factor
TLR2	Toll-like receptor 2
TSA	Trichostatin A
TrxG	Trithorax group
UTR	Untranslated region
VPA	Valproic acid
VWM	Vanishing white matter
Xist	X inactive transcript

Chapter 1
Reprogramming: A New Era in Regenerative Medicine

Abstract Embryonic stem cells (ESCs) exhibit the capacity for unlimited self-renewal and an ability to generate all somatic cell lines. However, political, ethical and practical obstacles, such as rejection of ESC-derived tissue by patients, obstruct the potential for using human ESCs (hESCs) in regenerative medicine. Still, the extreme plasticity and proliferative nature of ESCs make them the 'gold standard' to match or beat. While some reprogramming technologies, such as somatic cell nuclear transfer (SCNT), are capable of generating ESC-like states they face similar challenges associated with ESCs. In 2006, Takahashi and Yamanaka reported the development of so-called "induced pluripotent stem cells" (iPSCs) from adult mouse fibroblasts. These cells were produced by inducing the expression of four transcription factors (TFs). In the last few years, many alternative reprogramming strategies have been studied in order to develop a safe and efficient method for therapeutic applications.

Keywords Embryonic stem cells · Pluripotent stem cells · Reprogramming · (de)differentiation · Trans-differentiation · Neural stem cells

1.1 Stem Cells

For mammals, fertilization of the female egg begins a steady, organized process of cell growth and differentiation. Cells in the earliest developmental stages can generate both embryonic and extra-embryonic tissue, meaning the entire range of cellular material needed to form and nurture the developing organism (Kelly 1977). The potential of a cell to generate other cell types is known as "potency" and as such these earliest cells of the reproductive cycle are termed "totipotent". Still within early development, and prior to implantation, differentiation begins to coincide with a loss of potency. Within the inner cell mass (ICM) of the blastocyst, cells no longer have the ability to form extra-embryonic tissue. But they can generate all 3 germ layers of the developing embryo (ectoderm, mesoderm, and endoderm) and so an entire organism. These are termed "pluripotent". As the reproductive cycle continues, from implantation to birth, cell potency drops precipitously. Only a select number of mammalian cells retain potency, namely adult stem (SC) and progenitor (PC) cells, and are limited to tissue generation within specific lineages. SCs and PCs can undergo two types of divisions, (1) symmetric cell division to expand their own numbers, and (2) asymmetric cell division to renew themselves and to give rise to a more differentiated progeny. Depending on how many cell types they can generate they are termed "multipotent" or "unipotent". The vast majority of cells making up a mature organism are in a "non-potent" state of terminal differentiation.

While multipotent and unipotent cells within human bodies could be used in regenerative medicine, pluripotent cells offer the most useful source of therapeutic material, given their ability to generate all tissues of the body. Naturally occurring pluripotent cells have been isolated and cultured from mice and humans, most notable of these being: embryonal carcinoma cells (ECCs) from tumours in germ cells, embryonic stem cells (ESCs) from the ICM of blastocysts, epiblast-derived stem cells (EpiSCs) from embryos after implantation, embryonic germ cells (EGCs) from primordial germ cells of mid-gestation embryos, and germline stem cells (mouse adult germline stem cells, maGSCs) from mouse testicular tissue (Stadtfeld and Hochedlinger 2010). The claim for isolation of human adult germline stem cells (haGSCs) from human testicular tissue is currently controversial (Conaco et al. 2006; Conrad et al. 2008; Ko et al. 2010).

Ultimately, ESCs have stood out among naturally derived pluripotent cell lines. On top of generating all 3 embryonic germ lines, and exhibiting unlimited self-renewal and proliferation, ESCs can be used to form chimeras (including entire ESC animals). This stands in contrast to other pluripotent lines whose failure in developmental assays may be due to altered parental imprinting (Hochedlinger and Jaenisch 2006; Stadtfeld and Hochedlinger 2010). Of course chimeric and whole animal developmental assays are not available for human ESCs (hESCs) due to practical and legal limitations. It may seem plausible that they would perform the same way as mouse ESCs (mESCs), but this cannot be treated as factual. More importantly, using hESCs in regenerative medical therapy is associated with known risks. Without considerable stocks of hESCs and screening procedures to shift through them, patients would face similar tissue rejection issues as experienced for normal organ transplants. This is one of the problems regenerative medicine needs to overcome. A solution to this problem, and the focus of much ongoing research, is to create patient-specific ESC-like cells.

Three methods have been shown to produce ESC-like cells using fully mature cells or genetic material from them (Fig. 1.1). The process of shifting from a terminally differentiated state to ESC-like pluripotency is commonly referred to as "dedifferentiation". The oldest and best known technique, called somatic cell nuclear transfer (SCNT), involves placing genetic material from somatic cells into

1.1 Stem Cells

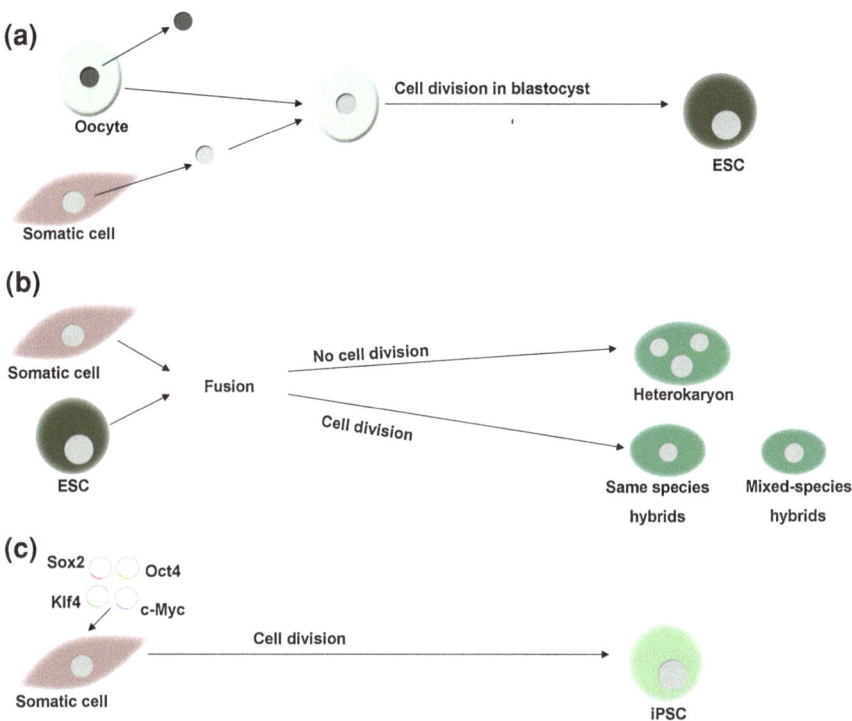

Fig. 1.1 De-differentiation of somatic cells. Overview of the methods used to produce pluripotent stem cells from differentiated cells. **a** Somatic nuclear transfer (SCNT) in which the nucleus of a differentiated cell is placed inside an enucleated oocyte. **b** Cell fusion (CF) involves the fusion of two different cell types which produce a hybrid if the nuclei merge, and heterokaryon cells containing multiple nuclei if they do not. **c** Pluripotency can be induced by viral transduction of the OSKM transcription factors, producing so-called iPSCs

enucleated oocytes. A more popular term for it is "cloning". SCNT has been used successfully for over half a century; beginning with amphibians (Briggs and King 1952; Gurdon 1962; Stadtfeld and Hochedlinger 2010) to reaching world headline status with the cloning of one specific mammal, "Dolly the sheep" (Wilmut et al. 2007). Many experiments used non-terminally differentiated cells, including embryonic tissue, as their source of genetic material. However, terminally differentiated sources have also proved successful with SCNT (Hochedlinger and Jaenisch 2006; Inoue et al. 2005; Stadtfeld and Hochedlinger 2010). In addition, fertilized eggs may be used instead of oocytes (Hochedlinger and Jaenisch 2006; Stadtfeld and Hochedlinger 2010). Cell fusion (CF) is another technique capable of creating ESC-like cells. In this case, different cell types are fused together; creating a single body that allows separate nuclei (and so genomes) to influence each other. If cell fusion products proliferate, the separate nuclei merge to form hybrids. Fusion products that do not proliferate and can retain separate nuclei are called heterokaryons. Heterokaryons hold an advantage over hybrids by preventing potential rearrangement or loss of chromosomes due to nucleic integration that would affect experimental conclusions. Inter-species heterokaryons provide still greater advantages as the source of transacting gene products can be clearly identified (Yamanaka and Blau 2010). Like SCNT, cell fusion has been studied for decades, with significant reversals in cell status seen as early as 1983 (Blau et al. 1983). The complete reprogramming of somatic cells came 14 years later, when they were hybridized with embryonic germ cells (Tada et al. 1997).

Practical and political considerations similar to those raised by the use of hESCs hinder SCNT and CF as regenerative therapies. However, both methods continue to be used to reveal the underlying mechanisms of cell differentiation and identity. SCNT experiments were the first to establish that cell differentiation is not the result of permanent changes in DNA sequence. If that were true, genetic material from fully differentiated cells could not give rise to all other cell types, especially entire living animals. Instead, "epigenetic" factors must work on, or in conjunction with, genetic material over the course of development to establish increasingly stable cell forms and functions. The results of CF work cited above added to this model, showing that while capable of producing stable and heritable cell states, epigenetic mechanisms allow sufficient plasticity for dramatic cellular changes. Particularly important for regenerative medicine, both kinds of experiments show that oocytes as well as fertilized eggs contain factors capable of reversing the epigenetic status of mature DNA, and so re-establish functional pluripotency. The final method of de-differentiation will be discussed in the next section "Reprogramming".

1.2 Reprogramming

Cellular reprogramming is the process in which one cell type is converted into another cell type. The Yamanaka lab was the first to reprogram somatic cells into pluripotent stem cells (Takahashi and Yamanaka 2006). This involved introducing 24 transcription factors (TFs) thought to maintain pluripotency into mouse fibroblasts using viral vectors (Takahashi and Yamanaka 2006). Afterwards, these factors were reduced and recombined to find the minimal factors needed to induce pluripotency. It turned out that only four factors were required: Oct4, Sox2, Klf4, and c-Myc (Takahashi and Yamanaka 2006). Afterwards the Yamanaka group also reported the successful reprogramming of human somatic cells into induced pluripotent stem cells (iPSCs) (Takahashi et al. 2007). Later studies identified that although Oct4 and Sox2 are crucial; the other two TFs can be left out or replaced by Nanog and LIN-28 (Nakagawa et al. 2008; Yu et al. 2007). However, the efficiency suffers significantly when reducing the number of TFs (Nakagawa et al. 2008).

The expression of Oct4 (octamer-binding transcription factor 4) is essential for the maintenance of a pluripotent state in ESCs (Amabile and Meissner 2009). Suppression of Oct4 as well as a two fold increase in expression leads to differentiation.

1.2 Reprogramming

Sox2 (sex-determining region Y-box 2) plays a role in the self-renewal of ESCs and forms a heterodimer complex with Oct4 (Amabile and Meissner 2009). Oct4 and Sox2 regulate the expression of several pluripotency-related TFs, including their own (Do and Scholer 2009). Oct4 and Sox2 also have an indirect influence on gene transcription by affecting chromatin structure, DNA methylation, microRNA and X chromosome inactivation (Do and Scholer 2009).

Klf4 (Krüppel-like factor 4) is a widespread TF with dual roles. Evidence suggests that Klf4 acts both as an oncoprotein and as a tumour suppressor, and that it can both activate and repress transcription, depending on the target gene and interaction partner (Amabile and Meissner 2009). A role as upstream regulator of Nanog, c-Myc and other core TFs has been proposed (Do and Scholer 2009). P53, an inhibitor of Nanog expression, is one of the targets inhibited by Klf4. Klf4 therefore stimulates the expression of Nanog. Another downstream effect of Klf4 is the activation of p21, which actually inhibits proliferation. However, p21 is in turn inhibited by c-Myc, indicating that the balance between Klf4 and c-Myc may be important when inducing pluripotency.

c-Myc plays a role in cell-cycle regulation, proliferation, growth, differentiation and metabolism (Amabile and Meissner 2009). c-Myc may induce global histone acetylation (Takahashi and Yamanaka 2006), which makes the DNA more accessible for TFs. iPSCs can be generated without c-Myc (Nakagawa et al. 2008), but this lowers the efficiency substantially. Since c-Myc is a known oncogene the use of c-Myc for therapeutic purposes should be as limited as possible.

Nanog (named after the mythological Celtic land of the ever-young 'Tir nan Og') was the first factor known to be involved in ESC self-renewal and pluripotency (Amabile and Meissner 2009). Loss of Nanog in ESC predisposes to differentiation. Surprisingly, Nanog is not an essential factor for iPSC generation, but this could be explained by the stimulation of Klf4 on endogenous Nanog expression.

LIN-28 is not a TF but an RNA-binding protein involved in developmental timing (Amabile and Meissner 2009). It acts as a translational enhancer by increasing the stability of specific messenger RNAs (mRNA).

The underlying mechanism by which reprogramming occurs is not well understood. It is thought that expression of the mentioned TFs leads to epigenetic modifications in the cells, such as chromatin and methylation changes, inducing a pluripotent state within the cells. After the transduction of these TFs it takes up to several weeks (see the review by Jaenisch and Young 2008) before a small proportion of the cells become iPSCs. It may be that the TFs gradually gain access to more DNA binding sites because of the hyperdynamics of the chromatin (Graf and Enver 2009). Furthermore, TFs related to ESCs have the ability to silence differentiation-affiliated TFs, leading to a suppression of the cells' original function (Graf and Enver 2009).

Over the last few years, alternative reprogramming techniques have been introduced and evolved to develop a method that is efficient and safe for therapeutic applications (Fig. 1.2). Below we will discuss these different reprogramming methods.

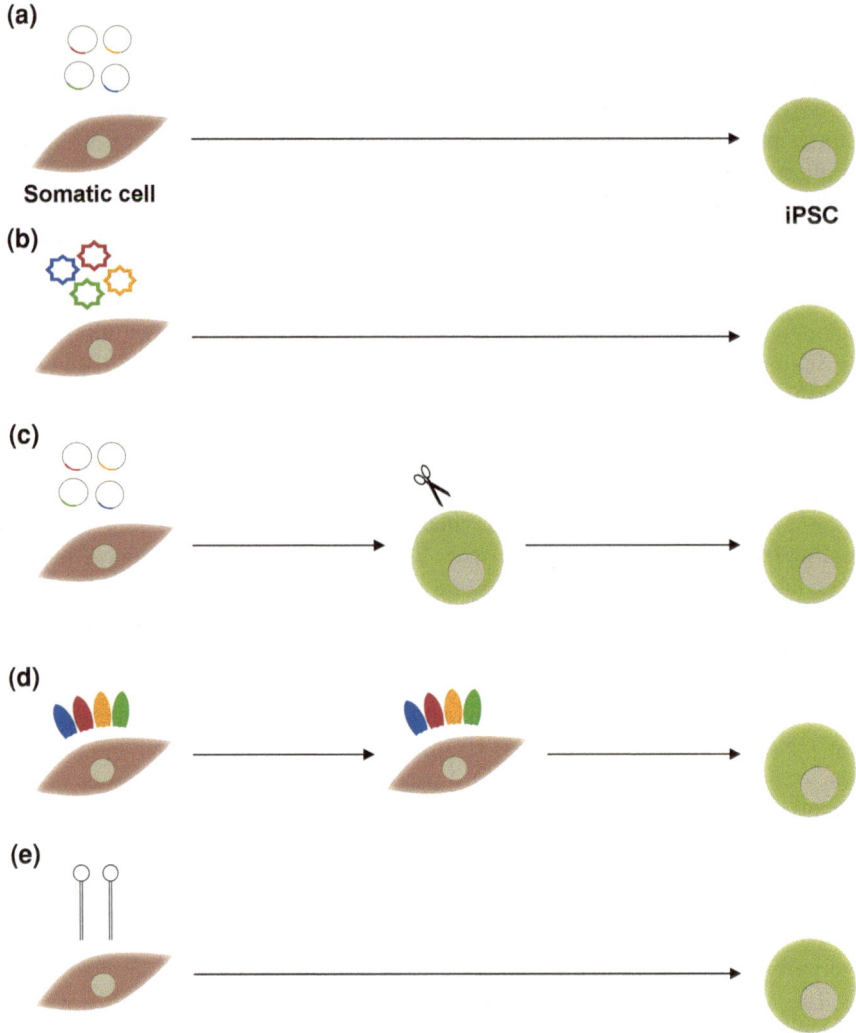

Fig. 1.2 Reprogramming methods. Overview of the methods used to reprogram somatic cells. **a** Integrating viral vectors. **b** Adenoviral associated vectors which do not integrate into the genome. **c** Excisable viral vectors, which are integrated into the genome but can later be excised. An example of this system is the Cre-LoxP system. **d** Repetitive rounds of the addition of recombinant cell-penetrating reprogramming proteins to the cells. **e** Transfection with miRNAs

1.2.1 Viral Vectors

The first reprogrammings were done with lentiviral or retroviral transduction. Although this has been one of the most efficient methods to produce iPSCs (Raya et al. 2009), there are some problems associated with this method. Both lentiviral and retroviral vectors integrate into the genome. A vector that integrates into the genome can disrupt the function of a gene at its site of entry, which can lead to undesirable effects due to a gain or loss of function (Clarke and van der Kooy 2009). One of the biggest associated problems is tumourigenesis.

Another problem is the expression of the TFs themselves. They are necessary for a period to establish a pluripotent cell line, but afterwards their expression needs to be shut down. A prolonged expression of the TFs can lead to problems with differentiation and especially the expression of the oncogenes c-Myc and Klf4 can be troubling. Continued expression of c-Myc during cell differentiation is related to tumour formation (Raya et al. 2009). In ESCs retroviral expression seems to be suppressed (Takahashi and Yamanaka 2006), and there is some evidence that it is suppressed in iPSCs as well (Raya et al. 2009). Hanna et al. (2007) did not observe any tumour formation in their animals after employing the iPSC technology, however, they caution against premature interpretation (Hanna et al. 2007). It is still possible that tumours form at a later time point. In addition, re-expression of c-Myc and Klf4 upon differentiation cannot be excluded.

Because of these problems with integrating viruses, different methods to induce pluripotency were developed. One of them is to use adenoviral vectors. Adenoviral vectors are somewhat safer due to their lack of genomic integration. Mouse liver cells and fibroblasts have been used to produce iPSCs by this method, although the efficiency is lower than with retroviral vectors and re-expression of the reprogramming factors is still a problem (Stadtfeld et al. 2008). Another study used Epstein-Barr virus-derived vectors to deliver the TFs. These vectors remain separate from the DNA of the host cell and are gradually lost with prolonged culturing, and vector-free cells can then be selected (Yu et al. 2009). Huangfu et al. (2008) showed that transduction of Oct4 and Sox2 together with valproic acid (VPA) treatment is sufficient to induce pluripotency in human fibroblasts (Huangfu et al. 2008). VPA is a histone deacetylase inhibitor, and therefore stimulates transcription and may make the DNA more accessible for Oct4 and Sox2.

Soldner et al. (2009) used the Cre/LoxP system to reprogram patient cells. They actually found a greater resemblance between the factor-free iPSCs and ESCs than to virus-carrying iPSCs. This indicates that the transgenes introduced in the iPSCs have an adverse effect on the cells' molecular properties. Further analyses revealed markers for pluripotency, the potential for multilineage differentiation and a normal karyotype in approximately 92% of all cell lines. However, the reprogramming efficiency was lower as compared to the iPSCs derived from retroviral transduction, and the Cre/LoxP system always leaves a small part of DNA behind, which might disrupt gene function.

A study by Woltjen et al. (2009) used a piggyBac transposon to deliver the TFs to the cells. Transposons are DNA sequences that have the ability to move over DNA strands with the help of transposase proteins. Transposases are responsible for cleaving and reinserting transposons into the genome. With this method, the transposon would first be integrated into the genome, allowing transcription of the reprogramming factors and the induction of pluripotency. When iPSC lines are stable and are ready to be differentiated again, a transposase can be delivered again to cut out the transposon. Because the transposons cleave the DNA seamlessly, this method will not leave any exogenous DNA behind. The drawback of this method is that transposases can both cleave and reinsert transposons. This means that integration in a different part of the genome cannot be completely avoided, unless there is a transposase available that lacks the pasting function. In addition, it still needs to be proven that this method will work for human cells and that removal of reprogramming genes is complete (Clarke and van der Kooy 2009).

1.2.2 Non-Viral Methods

A number of studies focused on inducing pluripotency without the use of viral vectors. It was shown that pluripotency can be induced with plasmid vectors. A study by Okita et al. (2008) generated murine iPSCs without the use of viral vectors. They transfected the cells with expression plasmids that only contained complementary DNAs of the four reprogramming factors, and did not find any genomic integration. On the contrary, a study by Kaji et al. (2009) using plasmid vectors found genomic integration, regardless of successful removal with the piggyBac transposon or the Cre-loxP system.

A number of small molecules were identified, that enhance the reprogramming process, and can be used to reduce the number of necessary reprogramming factors. These small molecules are thought to enhance activation of Oct4/Sox2 by epigenetic modifications (Do and Scholer 2009). A study by Zhou et al. (2009) showed that the Oct4, Sox2, Klf4 and c-Myc proteins can be delivered directly into the cell by fusing them with a small peptide that is cell-membrane permeable. They named these protein constructs recombinant cell-penetrating reprogramming proteins. This method needs multiple rounds of addition of recombinant proteins to make sure the proteins are present long enough for reprogramming to take place, so it is a labour intensive method. However, the advantage is that the proteins are broken down by the cell itself and nothing of the reprogramming constructs remains inside the cells.

Recent studies indicate great potential for reprogramming with microRNAs (miRNAs). miRNAs are regulatory RNAs that repress the expression of a large set of target genes post transcriptionally (Chang and Gregory 2011). A study by Anokye-Danso et al. (2011) showed that reprogramming can be done with miRNAs from the miR302/367 cluster. This cluster is known to be upregulated in ESCs. Reprogramming with these miRNAs was faster and approximately 100-fold more efficient than with the standard Oct4-Sox2-Klf4-c-Myc (OSKM) viral vectors; but the study of Anokye-Danso et al. (2011) still used viral vectors to deliver the miRNAs to the cell. This barrier was overcome in a study reporting the reprogramming of mouse and human fibroblasts by direct transfection of double stranded miR-200c, in combination with miR-320s, and 369s (Miyoshi et al. 2011). The length of reprogramming was similar to viral OSKM protocols, though the efficiency was quite low ($<0.002\%$). Together, these studies have proven the principle of using miRNAs as a reprogramming strategy, with the latter removing risks inherent to viral vectors (Sridharan and Plath 2011). However, more work is required to improve pure miRNA based protocols to make them more efficient and thus practical as a clinical technique.

1.3 Pluripotency, Efficiency, and Identity

1.3.1 Measuring Pluripotency

Regardless of reprogramming strategy, the endpoint of regenerative medical therapies will be partial-to-terminally differentiated cells. Therefore, methods are needed to evaluate the success and efficiency of reprogramming. In addition, for iPSC-based therapies that start with dedifferentiation, methods are required to determine the success and efficiency of reprogramming to an ESC-like state. Much work has been done on this latter point, and it elucidates the complexity of assessing, understanding, and changing cell identity (Stadtfeld and Hochedlinger 2010). Basic functional assays for iPSCs are the same as assays for successful derivation of competent ESCs, and can be distinguished in in vivo and in vitro assays.

Molecular assays have been created to indicate generation of a pluripotent, and so ESC-like, condition in vitro (Hanna et al. 2010). A commonly used assay is the detection of proteins or mRNAs associated with pluripotency, by either immunocytochemical stainings or reverse transcription polymerase chain reaction (RT-PCR) respectively. Alkaline phosphatase (AP) is often used as a marker of pluripotent cells. The epigenetic status of the cells can be determined through bisulfate genomic sequencing, luciferase reporter assays for promoter activity and chromatin immunoprecipation for histone modification status (Mikkelsen et al. 2008). The main drawback of these techniques is that in some somatic cells the expression or epigenetic status of pluripotency genes may be similar to pluripotent cells, making it harder to draw conclusions about the success of reprogramming. However, more in-depth epigenetic techniques, and testing other pluripotency genes might provide a solution.

In vivo, cells can be tested by injection into immunologically compromised mice (Stadtfeld and Hochedlinger 2010), with histological analysis of any resulting teratomas to determine the generation of ecto-, meso-, and endodermal tissue. The final test for pluripotent cells is the injection of labeled cells into ICM of unlabeled embryonic blastocysts, their transfer into a maternal host, and observation of subsequent offspring for chimeric properties including contribution to germline with second generation offspring exhibiting inherited traits (Stadtfeld and Hochedlinger 2010). This last assay can be tightened by injection of ESCs/iPSCs into tetraploid blastocysts that are inherently incapable of embryonic tissue development, resulting in all ESC/iPSC labeled offspring. Practical limitations to these functional assays, is that they require weeks to months to show results.

Also important for proposed medical use, practical and legal issues prevent the use of chimeric and tetraploid complementation assays for human ESCs and iPSCs.

1.3.2 Reprogramming Efficiency

Reprogramming efficiency, the percentage of viable iPSC colony forming cells generated by a given technique is influenced by various factors, but is generally very low (Stadtfeld and Hochedlinger 2010). One of these factors is the somatic cell source. The first iPSCs were derived from fibroblasts (Takahashi and Yamanaka 2006). However, a study by Kim et al. (2009) showed that neural stem cell pluripotency can be induced by Oct4 alone. Neural stem cells are not very useful for therapeutic purposes, because of the invasive methods to retrieve them. But there might be cells other than fibroblasts that are easily available and need less reprogramming factors. Given that different cell types may transcribe any TF at sufficient levels, protocols may be altered to fit the cell. In 2010, two articles separately reported successful induction of pluripotency in human T-cells, suggesting that blood samples alone might be sufficient, eliminating the need for biopsies (Seki et al. 2010; Staerk et al. 2010). Use of blood samples would allow for disease modelling on previously stored blood, including from patients that have already died.

In addition to cell source, reprogramming efficiency is often dependent on a combination of cell state and reprogramming strategy. The measured efficiency for iPSCs derived from terminally differentiated cell types ranges from 0.002 to 1.4%, with most lying below 0.1%. However efficiency has been reported as high as 4.5% using a drug-induced method that ensures equal and adequate TF expression across cells (Wernig et al. 2008). Efficiency is noticeably higher for donor cells that are less differentiated to begin with such as stem and progenitor cells, or when using factors that help cells switch lineage, ranging from 0.004 to 25%. Of note, methods using synthetic-mRNA produced efficiencies around 1%, compared to DNA-based methods; suggesting that viral, transposon, and plasmid vectors could be replaced by non-DNA approaches (Stadtfeld and Hochedlinger 2010). As mentioned earlier, a recent study by Anokye-Danso et al. (2011) showed that reprogramming with miRNAs increases the efficiency ∼100 fold compared to the OSKM viral vectors, with an efficiency of 80% after 8 days of reprogramming. However, use of miRNAs without a viral vector was found to be comparable to the extreme low end of viral OSKM efficiency (Miyoshi et al. 2011).

Consistently low efficiencies in early viral vector, TF-based reprogramming experiments raised the question if reprogramming can be achieved in all cells or just a subset of special cells. This is captured in two competing models of reprogramming known as the stochastic and the elite/deterministic models (Hanna et al. 2009; Yamanaka 2009). In the stochastic model, low efficiency is the result of numerous random events that halts or reverses reprogramming, but which can be overcome given sufficient processing time or mechanisms (Hanna et al. 2009). In the elite/deterministic model low efficiency is the direct result of having a limited number of special cells capable of being reprogrammed, though these may also be affected by random events similar to the stochastic model. Proponents of elite/deterministic models argue that, given such low experimental efficiencies, most cells are simply not reprogrammable (Yamanaka 2009). An experiment involving extended, continuous reprogramming showed efficiencies rise from 3 to 5% at two weeks to over 92% by 18 weeks (Hanna et al. 2009). That is strong evidence for the stochastic model, suggesting that all cells are capable of reprogramming. However, in a review article Stadtfeld and Hochedlinger (2010) suggest a combined elite/stochastic model based on as yet unpublished data from their lab supposedly showing that certain "refractory fibroblast populations" cannot be reprogrammed. In that model, most cells would be open to reprogramming with a select set unable to do so due to some extreme block to resetting epigenetic state. In either case, the existence of a stochastic element argues that efficiency might be improved by addressing possible rate limiting steps.

Hanna et al. (2009) also investigated the effects of expressed genes and culture factors on reprogramming rates. Increased expression of the TF Nanog was found to accelerate reprogramming in a cell-division independent manner, reducing required cell divisions to produce iPSCs from 70 to 50. This indicates an overcoming, perhaps reduction, of stochastic events blocking iPSC transitions. Alternatively, the overexpression of Lin28, or the inhibition of p53/p21, enhanced reprogramming in a cell division dependent manner, increasing cell divisions by 30% and dropping the required time for achieving 93% reprogramming efficiency from 17 to 8 weeks. This effect can be explained by rapid division creating more chances for positive stochastic events to occur, or that mechanisms involved in cell division inherently assist epigenetic reprogramming. However, it was noted that even these accelerated iPSC rates could not compete with SCNT rates, which require just one-two cell divisions to achieve reprogramming (Egli et al. 2007; Hanna et al. 2009). This is similar to results from cytoplasm-protein transfer that needed only 1 transfer of reprogramming material, with no prolonged, forced overexpression of pluripotency genes (Cho et al. 2008), suggesting that more direct, active reprogramming factors exist in embryonic cytoplasm than are employed using the limited TF-based reprogramming methods. It is plausible these as yet undiscovered epigenetic mechanisms might have been capable of handling whatever block Stadtfeld and Hochedlinger (2010) encountered in their unpublished work, which would work to establish a pure stochastic model recommended by Yamanaka (2009) and indicated by Hanna et al. (2009).

1.3.3 iPSCs Versus ESCs

Since the first papers about iPSCs there have been doubts about their claimed similarity to ESCs, regardless of extensive studies on ESC resemblance in appearance, surface antigens, gene expression, epigenetic status of pluripotent cell-specific genes, telomerase- and differentiation activity (Okita et al. 2008; Park et al. 2008; Soldner et al. 2009; Takahashi and Yamanaka 2006). In 2009, a controversial paper reported a series of meta-analytical comparisons between iPSCs and ESCs, finding a difference in gene expression regardless of species, cell type, and lab (Chin et al. 2009). Gene expression differences were significant and consistent enough for the authors to call iPSCs a separate subtype of pluripotent cells, even though miRNA levels only showed minor differences and epigenomic methylation was similar. That same year another study compared the gene expressions levels in which the possibility of viral vector influence had been removed (Marchetto et al. 2009). It was suggested that such differences might occur from a retained "memory" of the donor cell type. However this would seem to conflict with the first article in that one would have expected to see an observable difference based on cell type. Both agreed that potential functional differences between iPSCs and ESCs arising from such gene expression differences should be investigated.

This started a series of experiments concerned with the accurate determination of cellular identity, particularly with respect to gene expression levels and retention of epigenetic memory in iPSCs. One study found very few differences between mouse iPSCs and ESCs using wide-scale molecular assays, in conflict with Chin et al. (2009), but that these few were capable of generating important functional differences (Stadtfeld et al. 2010). Another difference from Chin et al. (2009) is that aberrant silencing was shown not to be a unique property inherent to mouse iPSCs (miPSCs), as some did express Dlk1-Dio3, and in any case silenced iPSCs could be rescued by enhanced reprogramming with an epigenetic HDAC inhibitor. A second article also found very small variations in gene expression and epigenetic signatures between human iPSCs (hiPSCs) and hESCs, but the authors felt these were not significant enough to differentiate them from one another, much less suggest functional differences (Guenther et al. 2010). Potential problems with the meta-analytical methods used by Chin et al. (2009) were pointed out by multiple papers (Guenther et al. 2010; Newman and Cooper 2010), and altered meta-analytical methods showed that gene expression patterns were tied to laboratories and not cell types. iPSCs may not have unique signatures as newer techniques and increased passaging clearly minimize differences with ESCs. But some differences are seen (even if not uniform) and results such as those mentioned earlier by Stadtfeld et al. (2010) prove that minute expression differences can have great functional consequences (Chin et al. 2010).

Therefore, there is an argument for having better and more unified standards of analyzing the genome wide molecular assay data. This debate was supported by Loh and Lim (2010) and illustrated by a study that followed several different cell lineages through continuous iPSC reprogramming (Polo et al. 2010). The authors found that iPSCs retain an epigenetic memory of their source cell type until sufficient passaging of the cells has occurred. Further they showed that epigenetic memories bias re-differentiation of iPSCs toward their previous lineages.

More recent studies, using greater analytical techniques, have continued to generate controversy over the status of iPSC fidelity to ESCs (Panopoulos et al.

2011). Of note is a study reporting the first whole genome DNA methylation profile created at single-base resolution (Lister et al. 2011). This group compared human somatic cells, ESCs, and five iPSC lines from different cell sources/reprogramming strategies, as well as somatic cells differentiated from the iPSCs. The authors state that at such fine scales iPSCs display aberrations in DNA methylation and histone modification patterns when compared to ESCs and their original cell type. Thus iPSC reprogramming involves both incomplete reprogramming as well as placement of unique epigenetic marks. Aberrations were generally located at telomeres and centromeres, indicating a possible physical impediment to reprogramming. Equally important was evidence suggesting that marks are retained despite passaging and differentiation. These results by Lister et al. (2011) echo, though at a much finer scale, those reported by Chin et al. (2009). Indeed, the authors argue that there may be a unique iPSC signature. Whether these results are confirmed over more cell lines, greater passage lengths, and greater extents of differentiation is yet to be seen. It is also plausible that localized, physical impediments to reprogramming can be overcome using epigenetic or chemical methods.

Ultimately, the extent of similarity between iPSCs and ESCs depends on a combination of reprogramming strategy, culturing conditions, and initial epigenetic status, with strong differentiation memories hindering de-differentiation and even faint memories (or improperly placed marks) biasing re-differentiation from putative iPSC states.

1.3.4 State of Pluripotency

As stated earlier, ESCs are the gold standard to match or beat for pluripotency. This suggests that ESCs are a specific type of near identical cells, which by their nature are stably pluripotent. In reality, while derived from a common location, the ICM of pre-implantation blastocysts, ESCs are cells whose developmental path includes a transient stage of pluripotency. This stage can be halted and preserved under proper isolation and culturing conditions. Next to that, not all species are equally permissive to ESC derivation.

Nichols and Smith (2009) studied characteristics of pluripotent cells derived from differentially permissive mouse strains, as well as between ESCs and EpiSCs. They found sufficient criteria to distinguish two flavors of pluripotency: naïve and primed (Nichols and Smith 2009). The naïve state is considered the ground state of pluripotency, which best represents the gold standard referred to earlier. The primed state, while pluripotent, appears set for differentiation and does not allow for some of the most stringent tests of pluripotency such as contribution to germline and tetraploid complementation. This may be due to having an inactivated X-chromosome, which, along with flat cell morphology, is a key distinguishing feature for primed cells. In mice, naïve and primed states primarily correspond to pluripotent ESC and EpiSCs, respectively. Thus, implantation drives priming of mouse blastocyst cells for differentiation, by pushing them away from a purely naïve pluripotent state. However, there are nonpermissive mouse strains, such as non-obese diabetic strains, whose ESC also exhibit EpiSC or primed characteristics.

This suggests that implantation itself may not be the causative factor, rather being coincidental, and that rate of development or amenability to culturing may be different between strains. Thus a cell taken from the ICM may be more likely to continue differentiation to a primed state in one strain than another. More importantly, and provocatively, it was pointed out that human ESCs closely resemble mouse EpiSCs across many criteria including culture conditions, gene expression patterns, X-chromosome inactivation, and morphology, and as such should be categorized as primed pluripotent along with mouse EpiSCs regardless of their derivation from the same developmental niche as mESCs (Nichols and Smith 2009, Hanna et al. 2010b). Natural differences between species and strains may very well prohibit derivation of naïve and so gold standard quality ESCs from all but permissive species and strains. Naturally the question was raised if altering methods, particularly culture conditions, could allow derivation of naïve ESCs from humans as well as other species. This state can be reached for hESCs and hiPSCs, with the transgenic expression of either Oct4 and Klf4, or Klf4 and Klf2, in a serum free culture medium containing leukaemia inhibitory factor (LIF), as well as inhibitors of the glycogen synthase kinase 3β (GSK3β) and extracellular signal-regulated kinases (ERK1/2) pathways (Hanna et al. 2010a). Continuous expression of TFs was required but cells could be passaged over 50 times. A method free of ectopic TF expression was also found, by supplementing the above culture medium with Forskolin, which induces endogenous expression of Klf4 and Klf2. However, these latter cells could not be maintained for more than 15–20 passages.

Cells generated by both methods fit many important criteria of naïve pluripotency found in mESCs, such as active X-chromosomes and domed morphology. Although there may be differences that have yet to be discerned, a proof of concept was established. In addition, another report indicated that X-chromosomal silencing common to hESCs was likely the product of derivation protocols, particularly exposure to atmospheric oxygen levels (Lengner et al. 2010). The authors were able to produce hESCs in a pre-X inactivation state, by switching to physiological oxygen concentrations, or introducing antioxidants to the cells. They further argued that other cell stress conditions beyond oxygen levels could drive X-inactivation, and so should be considered during hESC derivations. This indicates that even simple protocol modifications can have profound effects on resulting ESC and iPSC states. More comprehensive histories of the experiments establishing types of pluripotency and defining criteria for these states can be found in recent review articles (Hanna et al. 2010b; Buecker and Geijsen 2010).

Evidence suggests that cellular identity is basically plastic, regardless of species and type, and amenable to changes when experiencing appropriate internal and/or external conditions. Required conditions will differ based on genetic and epigenetic background of any given source cell, but no prohibitions are likely to exist regarding target cell identity, even ground state naïve pluripotency. However, better paths to reach these endpoints and more definitive landmarks of success need to be identified.

1.4 Differentiation

Regenerative medicine most often requires restoring tissue with a specialized cell type. Next to safe and efficient methods to induce pluripotency, optimized differentiation protocols are needed to produce the desired cell types. Moreover, transplantation of undifferentiated cells can lead to the formation of teratomas (Giudice and Trounson 2008). So cell transplant without pluripotent cells should be warranted. Most differentiation studies are done on ESCs. Theoretically, the differentiation of ESCs and iPSCs should be similar. Therefore, we will not make a distinction between the two pluripotent stem cell populations. However, small differences between them could influence differentiation into the different lineages (discussed in Sect. 1.5). Which factors are necessary for the differentiation of course differs for different cell types. Because of our focus on brain diseases, in this article the differentiation into neurons (more specifically motor and dopaminergic neurons), astrocytes and oligodendrocytes will be discussed.

During development of the central nervous system (CNS), pluripotent ESCs produce multipotent neural stem cells (NSCs), which in turn generate neurons and glial cells (astrocytes and oligodendrocytes). NSCs are thought to go through three stages of gestation (Fig. 1.3). Early on they are restricted to growth and self-renewal (Fujita 1986, 2003). This is followed by a stage allowing for asymmetric division into NSC and neuronal cells, or later in that stage NSC and neural precursor cells (NPCs) (Noctor et al. 2004). The NPCs are themselves limited to producing neurons, and so unipotent. Finally, as they reach maturity, NSCs become truly mutipotent, capable of producing all three brain cell types (Qian et al. 2000).

There are different protocols to generate NSCs from pluripotent SCs, but the following method is one of the most commonly used methods. Pluripotent SCs are grown on a layer of irradiated mouse embryonic fibroblasts (MEFs). For the differentiation into NSCs, the SCs are detached from the MEF layer and transferred into a suspension culture containing no growth factors, where they form embryoid bodies (EB). After a few days (5–8), the cells are changed to a tissue culture flask with a defined neural inducing medium (Zhou et al. 2008a). Different factors are added to promote the NSC state, for example noggin. After 5–10 days, the formation of a rosette-like shape that contains small elongated cells can be detected at the centre, surrounded by flattened cells. This rosette-like structure in the middle contains neuroepithelial cells, which can be easily selected because these cells only attach loosely to the surface and can be picked up by a pipette. These neuroepithelial cells can then be differentiated in any neuronal or glial cell type (Li et al. 2008; Zhang et al. 2001).

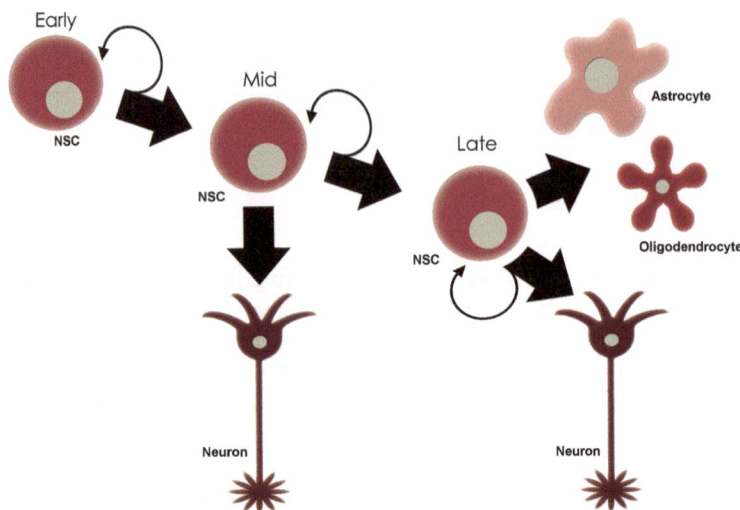

Fig. 1.3 CNS development. NSCs go through three stages that end in an ability to generate both neurons and glia

1.4.1 Motor Neurons

For the differentiation into motor neurons, neuroepithelial cells are cultured in retinoic acid (RA) and sonic hedgehog (SHH). RA is involved in the anterior-posterior patterning in early developmental stages and activates multiple genes involved in cell differentiation through binding to retinoic acid receptors. SHH activates the hedgehog signalling pathway, which plays a crucial role during development.

Differentiation into subtypes of neurons often yields mixed populations. The problem with this mixed population is not just that there is less of the cell type of interest, but also that it is not known what kind of cells are present in this population. Most studies find an efficiency of 20% for the differentiation into motor neurons (Li et al. 2008; Shin et al. 2007; Soundararajan et al. 2007). The common protocol involves adding RA and SHH, and, upon the first sign of Olig2-expressing progenitors the amount of SHH is 10-fold reduced. In the study of Li et al. (2008) it was found that a prolonged exposure to high concentrations of SHH led to a higher efficiency, namely 50%. They also found that 96% of the cells were either motor neurons or Olig2 expressing progenitors. For the development into motor neurons, first Olig2 expressing progenitors must be present. Then Ngn2 is expressed, leading to a suppression of Olig2 and a differentiation into motor neurons. When Ngn2 is not expressed, Olig2 expressing progenitors can develop into oligodendrocytes. The Olig2 expressing progenitors still present in the study of Li et al. (2008) may therefore develop into oligodendrocytes instead of motor neurons. However, this study at least identified that within the mixed population of cells generated by differentiation into motor neurons 96% consist of motor neurons, Olig2 expressing progenitors and in later stages, oligodendrocytes.

1.4.2 Dopaminergic Neurons

The differentiation into dopaminergic neurons can be mediated in different ways. The first studies used feeder layers of mice PA6 or MS5 cells. Adding certain growth factors to the medium, like fibroblast growth factor 8 (FGF-8), brain-derived neurotrophic factor (BDNF) and glial-derived neurotrophic factor (GDNF) or the anti-oxidant ascorbic acid (AA) increases the number of tyrosine hydroxylase (TH)-positive cells significantly (Grivennikov 2008; Yu et al. 2007). Other studies showed that the withdrawal of certain growth factors such as FGF-2 promoted neuronal differentiation (Zhang et al. 2001). TH is involved in the dopamine synthesis and often used as an early marker for dopaminergic neurons. An efficiency of about 20–40% TH-positive cells is often found (Hwang et al. 2010).

Other studies used co-culturing with human midbrain astrocytes, which led to about 28% of dopaminergic neurons (Hwang et al. 2010). Iacovitti et al. (2007) developed a method to differentiate stem cells into dopaminergic neurons without the use of feeder layers, since the use of animal derived substances can give problems when the neurons will be used in a clinical setting (Iacovitti et al. 2007). They cultured the neural progenitors with basic FGF and put the rosettes in an adherent cell culture and incubated them with dibutyryl cyclic adenosine $3',5'$monophosphate (dbcAMP) to differentiate into dopaminergic neurons. In these cultures, 60–80% of the cells differentiated into dopaminergic neurons.

Most protocols for differentiation into dopaminergic neurons take a long time. A problem associated with this is that most cells form highly branched processes in this period, which makes it almost impossible to harvest these cells without damaging them (Iacovitti et al. 2007). Cho et al. (2008) used so-called spherical neural masses (SNMs). These SNMs are a kind of neurospheres grown from neural progenitor cells. These SNMs are passaged a couple of times, and in each passage cells with a non-neuronal morphology were removed. The advantages of this method are that the cells can be passaged for a long time without losing their differentiation capability, no feeder cell layers are required, and the efficiency of the induction of TH-positive neurons from these SNMs was 86%. Moreover, most of these cells expressed markers of midbrain-dopaminergic neurons, which may enhance their use for Parkinson's disease therapy. Transplantation into a rat-model of Parkinson's disease led to a behavioural recovery and no tumour formation was observed. This method seems to provide a safe and efficient way to produce dopaminergic neurons. Consequent analyses, such as immunocytochemical and electrophysiological data should be carried out as well to reveal a succesful derivation of functional midbrain dopaminergic neurons. This would most likely result in a reduced number of the initial TH positive neurons.

1.4.3 Oligodendrocytes

When NSCs are cultured in vitro without any specific differentiation factor, glial cells and especially oligodendrocytes only appear after prolonged culturing. Oligodendrocytes derive from oligodendrocyte precursor cells (OPCs). A number of protocols to derive OPCs and oligodendrocytes in vitro have been developed.

Nistor et al. (2005) differentiated human ESCs into oligodendrocytes by culturing in glia restricted media (GRM), combined with 7 days RA exposure. The components of the GRM that influences the differentiation into oligodendrocyte-lineage are the growth factor insulin, the differentiation factor triiodothyroidin hormone, and the growth factors FGF and endothelial growth factor (EGF) (Nistor et al. 2005). Insulin and insulin-like growth factor 1 (IGF-1) are known to be involved in myelination and oligodendrocyte survival. Thyroid hormones promote differentiation and regulate the timing of differentiation, and FGF and EGF are known to extend proliferation of neural progenitor cells and promote glial cell differentiation (Nistor et al. 2005).

Other protocols involve exposure to T3, which is known to induce oligodendrocyte survival and differentiation, noggin, platelet-derived growth factor (PDGF) and fibroblast growth factor 2 (FGF2), and RA (Buchet and Baron-Van 2009). Noggin induces expression of Sox10, an essential TF for oligodendrocytic differentiation, so noggin treatment at specific stages enhances the development of oligodendrocytes. Noggin also inhibits bone morphogenetic protein 4 (BMP4) which stimulates neuron and astrocyte formation and thereby inhibits oligodendrocyte formation. Inhibiting BMP4 may therefore enhance the efficiency of oligodendrocytic differentiation (Izrael et al. 2007). When RA treatment is used together with a selection of oligodendrocytic-lineage cells, over 80% OPCs were derived (Izrael et al. 2007).

1.4.4 Astrocytes

The factors LIF and BMP2 were the first factors which were identified to promote astrocyte differentiation in vitro (Nakashima et al. 1999). LIF is a cytokine involved in cell growth and development. BMPs are involved in early development and known to induce astrocyte and neuronal formation from neuroepithelial cells. However different BMPs seem to be involved in the differentiation into astrocytes than into neurons (Chang et al. 2003). A study by Mi et al. (2001) showed that endothelial cells can introduce astrocyte differentiation by secretion of LIF (Mi et al. 2001). Chang et al. (2003) showed that medium that was derived from astrocyte-enriched cultures, so-called astrocyte conditioned medium (ACM), could differentiate precursor cells into astrocytes. The efficiency was 84%, compared with 7% GFAP-positive cells in a control culture. Glial fibrillary acidic protein (GFAP) is a known marker for astrocytes. The factors that they identified as promoters of astrocyte differentiation were LIF, ciliary neurotrophic factor

(CNTF), BMP4 and BMP6. So, although different BMPs are found in different studies, LIF and BMP appear to be the most important factors to induce astrocyte differentiation.

1.4.5 Selection of Differentiated Cells

A last step before transplantation would be selection of the differentiated cells. Although neuronal and glial cell types are quite easily recognizable because of their distinct morphology, this process will be shortly discussed.

Identification can be done by time-consuming analysis like fluorescence-activated cell sorting (FACS) or RT-PCR. These methods often lead to a significant loss of viable cells, and for example FACS depends on the presence of specific markers on the cell surface (Giudice and Trounson 2008), which are not always present. Differentiated cells can also be identified with the help of a reporter gene. This reporter gene consists of a drug-selectable or a fluorescent protein. By attaching this reporter gene to a cell-type or tissue specific promoter, differentiated cells can be selected. There are multiple ways to introduce these reporter genes into the cell, including viral vectors, nucleofection and homologous recombination (Giudice and Trounson 2008). Of course, these methods have the same disadvantages associated with the introduction of viral vectors or genomic integration discussed earlier. Because the marker is only meant for selection purposes, permanent genome integration would not be favourable, especially when the cells will be used in clinical settings.

1.5 Trans-Differentiation

One problem common to all iPSC techniques is that the newly de-differentiated cells must then be re-differentiated to a desired cell type. That requires additional time, material, and energy. A less intensive approach would be to reprogram tissue directly to the desired type, and skip pluripotency. This technique is called "trans-differentiation", and two recent articles demonstrated its wide potential for regenerative medicine (Nicholas and Kriegstein 2010; Vierbuchen et al. 2010). The first article describes an experiment using injections of an adenovirus, co-expressing a pool of TFs, to reprogram mouse pancreatic exocrine cells into insulin producing β-cells in vivo (Zhou et al. 2008b). Required factors were eventually narrowed to three: Ngn3, Pdx1, and Mafa, with only transient expression needed to induce trans-differentiation. The newly reprogrammed cells appeared identical to endogenous β-cells, and were shown to rescue diabetic mice suffering from hyperglycemia.

The second article covered an in vitro trans-differentiation of mouse embryonic and postnatal tail-tip fibroblasts into functioning neurons (Vierbuchen et al. 2010).

For this experiment, lentiviruses were used to introduce TFs for reprogramming. It was found that of 19 possible factors, only three were required: Ascl1, Brn2, and Myt1l. Trans-differentiation was faster than regular iPSC reprogramming, requiring only 8 days, and under extensive testing the induced neurons exhibited proper function, most importantly producing action potentials and synapses. This second experiment was significant for showing trans-differentiation between cell lineages, which the first did not address. Together, these articles provide evidence for the in vivo and in vitro tissue generation capabilities desired for regenerative medicine, with less waste compared to iPSC methods.

Another article, by Kim et al. (2011), showed that transient expression of the OSKM reprogramming factors can trans-differentiate fibroblasts to NPCs when cultured in neural reprogramming medium. More recently it was reported that transfection of human fibroblasts with miR-124 and just two TFs (*BRN2* and *MYT1L*) via lentoviral vector, was capable of generating functional neurons (Ambasudhan et al. 2011). Positive signs of conversion were seen as early as 3 days, and significant numbers (46% positive for *NeuN*, a marker for neurons) at 15 days. Another study using lentiviral co-transduction of the TFs Ascl1, Brn2, and Zic1 in a neuron selective media obtained higher rates of direct conversion of human fibroblasts to neurons (Qiang et al. 2011). At 21 days $62 \pm 6\%$ of cells showed neuronal morphology as well as expression of appropriate gene markers. This rate was increased to $85 \pm 15\%$ with co-transduction of an additional TF: Myt1l. Qiang et al. (2011) additionally reported use of this technique on skin samples from patients with Alzheimer disease, in an early attempt at in vitro disease modelling using trans-differentiation rather than de-differentiation to pluripotency followed by re-differentiation.

It remains to be seen whether de- and trans-differentiation will play complementary roles, or if the latter will eclipse the former. Given the use of viral vectors and TFs, it is possible that safety and/or practical issues will drive efforts to find alternative reprogramming mechanisms for trans-differentiation, as it has for iPSCs. This possibility is supported by Liu et al. (2010). Rather than creating neurons (brain grey matter), the goal was to induce the formation of oligodendrocytes (brain white matter) from mouse fibroblast (Liu et al. 2010). Surprisingly, chemical compounds simply added to ease TF-mediated reprogramming induced the endogenous expression of myelin genes, indicating a shift away from fibroblast, while oligodendritic TFs alone could not. A combination of TFs and the chemical compounds did not significantly improve results, suggesting that the TFs were not required for the process. While oligodendrocytes were not generated, it is possible the short time scale for reprogramming may have been a factor. Also, potential enhancers for this specific trans-differentiation were identified, though not used in the experiment.

Trans-differentiation is in need of more research. Especially for regenerative medicine, trans-differentiation might be favoured over reprogramming followed by differentiation. First of all trans-differentiation might be faster, but more important, a transplantation with trans-differentiated cells will lower the risk of transplanting undifferentiated cells. Since undifferentiated cells might develop tumours, this is an important feature to take into account.

References

Amabile G, Meissner A (2009) Induced pluripotent stem cells: current progress and potential for regenerative medicine. Trends Mol Med 15:59–68

Ambasudhan R, Talantova M, Coleman R, Yuan X, Zhu S, Lipton SA, Ding S (2011) Direct reprogramming of adult human fibroblasts to functional neurons under defined conditions. Cell Stem Cell 9:113–118

Anokye-Danso F, Trivedi CM, Juhr D, Gupta M, Cui Z, Tian Y, Zhang Y, Yang W, Gruber PJ, Epstein JA, Morrisey EE (2011) Highly efficient miRNA-mediated reprogramming of mouse and human somatic cells to pluripotency. Cell Stem Cell 8:376–388

Blau HM, Chiu CP, Webster C (1983) Cytoplasmic activation of human nuclear genes in stable heterocaryons. Cell 32:1171–1180

Briggs R, King TJ (1952) Transplantation of living nuclei from blastula cells into enucleated frogs' eggs. Proc Natl Acad Sci U S A 38:455–463

Buchet D, Baron-Van EA (2009) In search of human oligodendroglia for myelin repair. Neurosci Lett 456:112–119

Buecker C, Geijsen N (2010) Different flavors of pluripotency, molecular mechanisms, and practical implications. Cell Stem Cell 7:559–564

Chang HM, Gregory RI (2011) MicroRNAs and reprogramming. Nat Biotechnol 29:499–500

Chang MY, Son H, Lee YS, Lee SH (2003) Neurons and astrocytes secrete factors that cause stem cells to differentiate into neurons and astrocytes, respectively. Mol Cell Neurosci 23:414–426

Chin MH, Mason MJ, Xie W, Volinia S, Singer M, Peterson C, Ambartsumyan G, Aimiuwu O, Richter L, Zhang J, Khvorostov I, Ott V, Grunstein M, Lavon N, Benvenisty N, Croce CM, Clark AT, Baxter T, Pyle AD, Teitell MA, Pelegrini M, Plath K, Lowry WE (2009) Induced pluripotent stem cells and embryonic stem cells are distinguished by gene expression signatures. Cell Stem Cell 5:111–123

Chin MH, Pellegrini M, Plath K, Lowry WE (2010) Molecular analyses of human induced pluripotent stem cells and embryonic stem cells. Cell Stem Cell 7:263–269

Cho MS, Lee YE, Kim JY, Chung S, Cho YH, Kim DS, Kang SM, Lee H, Kim MH, Kim JH, Leem JW, Oh SK, Choi YM, Hwang DY, Chang JW, Kim DW (2008) Highly efficient and large-scale generation of functional dopamine neurons from human embryonic stem cells. Proc Natl Acad Sci U S A 105:3392–3397

Clarke L, van der Kooy D (2009) A safer stem cell: inducing pluripotency. Nat Med 15:1001–1002

Conaco C, Otto S, Han JJ, Mandel G (2006) Reciprocal actions of REST and a microRNA promote neuronal identity. Proc Natl Acad Sci U S A 103:2422–2427

Conrad S, Renninger M, Hennenlotter J, Wiesner T, Just L, Bonin M, Aicher W, Buhring HJ, Mattheus U, Mack A, Wagner HJ, Minger S, Matzkies M, Reppel M, Hescheler J, Sievert KD, Stenzl A, Skutella T (2008) Generation of pluripotent stem cells from adult human testis. Nature 456:344–349

Do JT, Scholer HR (2009) Regulatory circuits underlying pluripotency and reprogramming. Trends Pharmacol Sci 30:296–302

Egli D, Rosains J, Birkhoff G, Eggan K (2007) Developmental reprogramming after chromosome transfer into mitotic mouse zygotes. Nature 447:679–685

Fujita S (2003) The discovery of the matrix cell, the identification of the multipotent neural stem cell and the development of the central nervous system. Cell Struct Funct 28:205–228

Fujita S (1986) Transitory differentiation of matrix cells and its functional role in the morphogenesis of the developing vertebrate CNS. Curr Top Dev Biol 20:223–242

Giudice A, Trounson A (2008) Genetic modification of human embryonic stem cells for derivation of target cells. Cell Stem Cell 2:422–433

Graf T, Enver T (2009) Forcing cells to change lineages. Nature 462:587–594

Grivennikov IA (2008) Embryonic stem cells and the problem of directed differentiation. Biochemistry (Mosc.) 73:1438–1452

Guenther MG, Frampton GM, Soldner F, Hockemeyer D, Mitalipova M, Jaenisch R, Young RA (2010) Chromatin structure and gene expression programs of human embryonic and induced pluripotent stem cells. Cell Stem Cell 7:249–257

Gurdon JB (1962) The developmental capacity of nuclei taken from intestinal epithelium cells of feeding tadpoles. J Embryol Exp Morphol 10:622–640

Hanna J, Cheng AW, Saha K, Kim J, Lengner CJ, Soldner F, Cassady JP, Muffat J, Carey BW, Jaenisch R (2010a) Human embryonic stem cells with biological and epigenetic characteristics similar to those of mouse ESCs. Proc Natl Acad Sci U S A 107:9222–9227

Hanna J, Saha K, Jaenisch R (2010b) Pluripotency and cellular reprogramming: facts, hypotheses, unresolved issues. Cell 143:508–525

Hanna J, Saha K, Pando B, van ZJ, Lengner CJ, Creyghton MP, van OA, Jaenisch R (2009) Direct cell reprogramming is a stochastic process amenable to acceleration. Nature 462:595–601

Hanna J, Wernig M, Markoulaki S, Sun CW, Meissner A, Cassady JP, Beard C, Brambrink T, Wu LC, Townes TM, Jaenisch R (2007) Treatment of sickle cell anemia mouse model with iPS cells generated from autologous skin. Science 318:1920–1923

Hochedlinger K, Jaenisch R (2006) Nuclear reprogramming and pluripotency. Nature 441:1061–1067

Huangfu D, Maehr R, Guo W, Eijkelenboom A, Snitow M, Chen AE, Melton DA (2008) Induction of pluripotent stem cells by defined factors is greatly improved by small-molecule compounds. Nat Biotechnol 26:795–797

Hwang DY, Kim DS, Kim DW (2010) Human ES and iPS cells as cell sources for the treatment of Parkinson's disease: current state and problems. J Cell Biochem 109:292–301

Iacovitti L, Donaldson AE, Marshall CE, Suon S, Yang M (2007) A protocol for the differentiation of human embryonic stem cells into dopaminergic neurons using only chemically defined human additives: Studies in vitro and in vivo. Brain Res 1127:19–25

Inoue K, Wakao H, Ogonuki N, Miki H, Seino K, Nambu-Wakao R, Noda S, Miyoshi H, Koseki H, Taniguchi M, Ogura A (2005) Generation of cloned mice by direct nuclear transfer from natural killer T cells. Curr Biol 15:1114–1118

Izrael M, Zhang P, Kaufman R, Shinder V, Ella R, Amit M, Itskovitz-Eldor J, Chebath J, Revel M (2007) Human oligodendrocytes derived from embryonic stem cells: Effect of noggin on phenotypic differentiation in vitro and on myelination in vivo. Mol Cell Neurosci 34:310–323

Jaenisch R, Young R (2008) Stem cells, the molecular circuitry of pluripotency and nuclear reprogramming. Cell 132:567–582

Kaji K, Norrby K, Paca A, Mileikovsky M, Mohseni P, Woltjen K (2009) Virus-free induction of pluripotency and subsequent excision of reprogramming factors. Nature 458:771–775

Kelly SJ (1977) Studies of the developmental potential of 4- and 8-cell stage mouse blastomeres. J Exp Zool 200:365–376

Kim J, Efe JA, Zhu S, Talantova M, Yuan X, Wang S, Lipton SA, Zhang K, Ding S (2011) Direct reprogramming of mouse fibroblasts to neural progenitors. Proc Natl Acad Sci U S A 108:7838–7843

Kim JB, Sebastiano V, Wu G, rauzo-Bravo MJ, Sasse P, Gentile L, Ko K, Ruau D, Ehrich M, van den BD, Meyer J, Hubner K, Bernemann C, Ortmeier C, Zenke M, Fleischmann BK, Zaehres H, Scholer HR (2009) Oct4-induced pluripotency in adult neural stem cells. Cell 136:411–419

Ko K, Rauzo-Bravo MJ, Tapia N, Kim J, Lin Q, Bernemann C, Han DW, Gentile L, Reinhardt P, Greber B, Schneider RK, Kliesch S, Zenke M, Scholer HR (2010) Human adult germline stem cells in question. Nature 465:E1

References

Lengner CJ, Gimelbrant AA, Erwin JA, Cheng AW, Guenther MG, Welstead GG, Alagappan R, Frampton GM, Xu P, Muffat J, Santagata S, Powers D, Barrett CB, Young RA, Lee JT, Jaenisch R, Mitalipova M (2010) Derivation of pre-X inactivation human embryonic stem cells under physiological oxygen concentrations. Cell 141:872–883

Li XJ, Hu BY, Jones SA, Zhang YS, Lavaute T, Du ZW, Zhang SC (2008) Directed differentiation of ventral spinal progenitors and motor neurons from human embryonic stem cells by small molecules. Stem Cells 26:886–893

Lister R, Pelizzola M, Kida YS, Hawkins RD, Nery JR, Hon G, Antosiewicz-Bourget J, O'Malley R, Castanon R, Klugman S, Downes M, Yu R, Stewart R, Ren B, Thomson JA, Evans RM, Ecker JR (2011) Hotspots of aberrant epigenomic reprogramming in human induced pluripotent stem cells. Nature 471:68–73

Liu J, Sandoval J, Doh ST, Cai L, Lopez-Rodas G, Casaccia P (2010) Epigenetic modifiers are necessary but not sufficient for reprogramming non-myelinating cells into myelin gene-expressing cells. PLoS One 5:e13023

Loh KM, Lim B (2010) Recreating pluripotency? Cell Stem Cell 7:137–139

Marchetto MC, Yeo GW, Kainohana O, Marsala M, Gage FH, Muotri AR (2009) Transcriptional signature and memory retention of human-induced pluripotent stem cells. PLoS One 4:e7076

Mi H, Haeberle H, Barres BA (2001) Induction of astrocyte differentiation by endothelial cells. J Neurosci 21:1538–1547

Mikkelsen TS, Hanna J, Zhang X, Ku M, Wernig M, Schorderet P, Bernstein BE, Jaenisch R, Lander ES, Meissner A (2008) Dissecting direct reprogramming through integrative genomic analysis. Nature 454:49–55

Miyoshi N, Ishii H, Nagano H, Haraguchi N, Dewi DL, Kano Y, Nishikawa S, Tanemura M, Mimori K, Tanaka F, Saito T, Nishimura J, Takemasa I, Mizushima T, Ikeda M, Yamamoto H, Sekimoto M, Doki Y, Mori M (2011) Reprogramming of mouse and human cells to pluripotency using mature microRNAs. Cell Stem Cell 8:633–638

Nakagawa M, Koyanagi M, Tanabe K, Takahashi K, Ichisaka T, Aoi T, Okita K, Mochiduki Y, Takizawa N, Yamanaka S (2008) Generation of induced pluripotent stem cells without Myc from mouse and human fibroblasts. Nat Biotechnol 26:101–106

Nakashima K, Yanagisawa M, Arakawa H, Taga T (1999) Astrocyte differentiation mediated by LIF in cooperation with BMP2. FEBS Lett 457:43–46

Noctor SC, Martinez-Cerdeno V, Ivic L, Kriegstein AR (2004) Cortical neurons arise in symmetric and asymmetric division zones and migrate through specific phases. Nat Neurosci 7:136–144

Newman AM, Cooper JB (2010) Lab-specific gene expression signatures in pluripotent stem cells. Cell Stem Cell 7:258–262

Nicholas CR, Kriegstein AR (2010) Regenerative medicine: cell reprogramming gets direct. Nature 463:1031–1032

Nichols J, Smith A (2009) Naïve and primed pluripotent states. Cell Stem Cell 4:487–492

Nistor GI, Totoiu MO, Haque N, Carpenter MK, Keirstead HS (2005) Human embryonic stem cells differentiate into oligodendrocytes in high purity and myelinate after spinal cord transplantation. Glia 49:385–396

Okita K, Nakagawa M, Hyenjong H, Ichisaka T, Yamanaka S (2008) Generation of mouse induced pluripotent stem cells without viral vectors. Science 322:949–953

Panopoulos AD, Ruiz S, Izpisua Belmonte JC (2011) iPSCs: induced back to controversy. Cell Stem Cell 8:347–348

Park IH, Zhao R, West JA, Yabuuchi A, Huo H, Ince TA, Lerou PH, Lensch MW, Daley GQ (2008) Reprogramming of human somatic cells to pluripotency with defined factors. Nature 451:141–146

Polo JM, Liu S, Figueroa ME, Kulalert W, Eminli S, Tan KY, Apostolou E, Stadtfeld M, Li Y, Shioda T, Natesan S, Wagers AJ, Melnick A, Evans T, Hochedlinger K (2010) Cell type of origin influences the molecular and functional properties of mouse induced pluripotent stem cells. Nat Biotechnol 28:848–855

Qian X, Shen Q, Goderie SK, He W, Capela A, Davis AA, Temple S (2000) Timing of CNS cell generation: a programmed sequence of neuron and glial cell production from isolated murine cortical stem cells. Neuron 28:69–80

Qiang L, Fujita R, Yamashita T, Angulo S, Rhinn H, Rhee D, Doege C, Chau L, Aubry L, Vanti WB, Moreno H, Abeliovich A (2011) Directed conversion of Alzheimer's disease patient skin fibroblasts into functional neurons. Cell 146:359–371

Raya A, Rodriguez-Piza I, Guenechea G, Vassena R, Navarro S, Barrero MJ, Consiglio A, Castella M, Rio P, Sleep E, Gonzalez F, Tiscornia G, Garreta E, Aasen T, Veiga A, Verma IM, Surralles J, Bueren J, Izpisua Belmonte JC (2009) Disease-corrected haematopoietic progenitors from Fanconi anaemia induced pluripotent stem cells. Nature 460:53–59

Seki Y, Kurisaki A, Watanabe-Susaki K, Nakajima Y, Nakanishi M, Arai Y, Shiota K, Sugino H, Asashima M (2010) TIF1beta regulates the pluripotency of embryonic stem cells in a phosphorylation-dependent manner. Proc Natl Acad Sci U S A 107:10926–10931

Shin S, Xue H, Mattson MP, Rao MS (2007) Stage-dependent Olig2 expression in motor neurons and oligodendrocytes differentiated from embryonic stem cells. Stem Cells Dev 16:131–141

Soldner F, Hockemeyer D, Beard C, Gao Q, Bell GW, Cook EG, Hargus G, Blak A, Cooper O, Mitalipova M, Isacson O, Jaenisch R (2009) Parkinson's disease patient-derived induced pluripotent stem cells free of viral reprogramming factors. Cell 136:964–977

Soundararajan P, Lindsey BW, Leopold C, Rafuse VF (2007) Easy and rapid differentiation of embryonic stem cells into functional motoneurons using sonic hedgehog-producing cells. Stem Cells 25:1697–1706

Sridharan R, Plath K (2011) Small RNAs loom large during reprogramming. Cell Stem Cell 8:599–601

Stadtfeld M, Apostolou E, Akutsu H, Fukuda A, Follett P, Natesan S, Kono T, Shioda T, Hochedlinger K (2010) Aberrant silencing of imprinted genes on chromosome 12qF1 in mouse induced pluripotent stem cells. Nature 465:175–181

Stadtfeld M, Hochedlinger K (2010) Induced pluripotency: history, mechanisms, and applications. Genes Dev 24:2239–2263

Stadtfeld M, Nagaya M, Utikal J, Weir G, Hochedlinger K (2008) Induced pluripotent stem cells generated without viral integration. Science 322:945–949

Staerk J, Dawlaty MM, Gao Q, Maetzel D, Hanna J, Sommer CA, Mostoslavsky G, Jaenisch R (2010) Reprogramming of human peripheral blood cells to induced pluripotent stem cells. Cell Stem Cell 7:20–24

Tada M, Tada T, Lefebvre L, Barton SC, Surani MA (1997) Embryonic germ cells induce epigenetic reprogramming of somatic nucleus in hybrid cells. EMBO J 16:6510–6520

Takahashi K, Tanabe K, Ohnuki M, Narita M, Ichisaka T, Tomoda K, Yamanaka S (2007) Induction of pluripotent stem cells from adult human fibroblasts by defined factors. Cell 131:861–872

Takahashi K, Yamanaka S (2006) Induction of pluripotent stem cells from mouse embryonic and adult fibroblast cultures by defined factors. Cell 126:663–676

Vierbuchen T, Ostermeier A, Pang ZP, Kokubu Y, Sudhof TC, Wernig M (2010) Direct conversion of fibroblasts to functional neurons by defined factors. Nature 463:1035–1041

Wernig M, Lengner CJ, Hanna J, Lodato MA, Steine E, Foreman R, Staerk J, Markoulaki S, Jaenisch R (2008) A drug-inducible transgenic system for direct reprogramming of multiple somatic cell types. Nat Biotechnol 26:916–924

Wilmut I, Schnieke AE, McWhir J, Kind AJ, Campbell KH (2007) Viable offspring derived from fetal and adult mammalian cells. Cloning Stem Cells 9:3–7

Woltjen K, Michael IP, Mohseni P, Desai R, Mileikovsky M, Hamalainen R, Cowling R, Wang W, Liu P, Gertsenstein M, Kaji K, Sung HK, Nagy A (2009) piggyBac transposition reprograms fibroblasts to induced pluripotent stem cells. Nature 458:766–770

Yamanaka S (2009) Elite and stochastic models for induced pluripotent stem cell generation. Nature 460:49–52

Yamanaka S, Blau HM (2010) Nuclear reprogramming to a pluripotent state by three approaches. Nature 465:704–712

References

Yu J, Hu K, Smuga-Otto K, Tian S, Stewart R, Slukvin II, Thomson JA (2009) Human induced pluripotent stem cells free of vector and transgene sequences. Science 324:797–801

Yu J, Vodyanik MA, Smuga-Otto K, Antosiewicz-Bourget J, Frane JL, Tian S, Nie J, Jonsdottir GA, Ruotti V, Stewart R, Slukvin II, Thomson JA (2007) Induced pluripotent stem cell lines derived from human somatic cells. Science 318:1917–1920

Zhang SC, Wernig M, Duncan ID, Brustle O, Thomson JA (2001) In vitro differentiation of transplantable neural precursors from human embryonic stem cells. Nat Biotechnol 19:1129–1133

Zhou H, Wu S, Joo JY, Zhu S, Han DW, Lin T, Trauger S, Bien G, Yao S, Zhu Y, Siuzdak G, Scholer HR, Duan L, Ding S (2009) Generation of induced pluripotent stem cells using recombinant proteins. Cell Stem Cell 4:381–384

Zhou JM, Chu JX, Chen XJ (2008a) An improved protocol that induces human embryonic stem cells to differentiate into neural cells in vitro. Cell Biol Int 32:80–85

Zhou Q, Brown J, Kanarek A, Rajagopal J, Melton DA (2008b) In vivo reprogramming of adult pancreatic exocrine cells to beta-cells. Nature 455:627–632

Chapter 2
Understanding Epigenetic Memory is the Key to Successful Reprogramming

Abstract Molecular biologists have developed powerful tools for measuring and/or describing a cell's epigenome, including: bisulfite-sequencing, chromatin-immunoprecipitation (ChIP), and comprehensive high-throughput arrays for relative methylation (CHARM). Over the last few years, a body of work has emerged using such techniques to map the entire epigenetic landscape of ESCs, iPSCs, and terminally differentiated cells. In this way the underlying mechanisms of differentiation, as well as the effects of artificial de-, re-, and trans-differentiation can be understood by comparing important differences in their epigenetic states.

Keywords Epigenetics · DNA methylation · Histone modification · Noncoding RNA · Epigenetic maps · CNS development

2.1 Epigenetic Influences on Gene Expression

Long term continuous TF-based reprogramming of iPSCs is required to achieve sufficient numerical efficiency as well as qualitative pluripotency. This suggests that TFs on their own face difficulties in rearranging epigenetic marks. This is somewhat logical, as most TFs would employ indirect mechanisms to adjust epigenetic states. Further, current repressive marks might have to be overcome for TFs to bind properly.

Different studies have shown that direct control of epigenetic mechanisms can improve reprogramming speed, efficiency, and quality. However, before discussing epigenetic factors that might aid reprogramming, a short overview of epigenetic mechanisms will be presented.

2.1.1 DNA Methylation

The most common form of epigenetic DNA modification is methylation of cytosine. Within mammals, and more prominently in humans, cytosine-guanine dinucleotide sequences tend to be grouped together in so-called CpG islands around gene promoters. Depending on frequency, this allows genes to be identified as having high (HCP) or low (LCP) CpG density promoters. Methyl groups are added to the cytosine residues of CpGs by various DNA methyltransferases (Dnmts). For example, Dnmt1 provides continuity of methylation patterns as a *maintenance* methyltransferase, while Dnmt3a and Dnmt3b establish new methyl groups as *de novo* methyltransferases. Once placed, methyl groups at promoter sites may prevent binding of transcription factors and so gene expression. They may also attract and assist repression complexes, which also prevent gene expression. This means that in general, methylation of DNA is a repressive epigenetic mark. Loss of Dnmts can cause severe deformities and death, indicating that proper epigenetic DNA methylation patterns are essential to development and health. One study determined that reaching a state of terminal differentiation requires Dnmt1 in specific, clarifying its importance to establishing cell identity (Jackson et al. 2004).

2.1.2 Histone Modification

Amino acid residues making up the tails of histone proteins are targets for modification. Such modifications include methylation, acetylation, phosphorylation, and ubiquitination. The specific effects of histone modification are dependent on position/type of amino acid, and type/degree of modification. Research is slowly building a histone code to relate modifications to epigenetic effects. This is a very large field of research and for practical reasons will not be discussed in detail. The most common, and therefore well researched, modifications are methylation and acetylation. Methyl groups are added by histone methyl-transferases (HMTs), while acetyl groups are added by histone acetyl-transferases (HATs). Histone deacetylation and demethylation is performed by HDACs and HDMs, respectively. There are also histone-remodeling complexes such as SWI/SNF capable of moving, eliminating, or swapping out specific histones which *de facto* removes present histone coding.

Acetyl groups are thought to open chromatin structure directly by introduction of a negative charge, while methyl groups tighten chromatin structure, directly or through interaction with protein complexes capable of structural/functional changes. Thus acetylation tends to be an activation mark, while methylation tends to be a repressive mark. It must be remembered though; active and repressive effects depend on many factors. For example, while tri-methylation of the 27th lysine residue of histone protein 3 (H3K27me3) is a repressive mark, trimethylation of the

2.1 Epigenetic Influences on Gene Expression

4th lysine of the same histone protein (H3K4me3) is an active mark. It is also possible to have more than one kind of modification on a single histone protein, and across histones of different nucleosomes. The polycomb group (PcG) of proteins contains important repressive complexes for gene regulation, particularly the HMT polycomb repressive complex (PRC)-2 that methylates H3K27, and HUT PRC1 that ubiquinates H2AK119. Problems with PRC2, including loss of subunits, have been shown to cause severe defects in embryonic development (Surface et al. 2010). In contrast, the Trithorax group (TrxG) of proteins includes HMTs that are equally important in placing active H3K4me3 marks. H3K4me3 is highly associated with DNA hypomethylation and upregulated gene expression. Interestingly, loss of both H3K27 and H3K4 trimethylation is correlated with DNA hypermethylation (Meissner et al. 2008). This indicates that epigenetic marking of histones, as DNA methylation, is important to proper development and health.

2.1.3 Noncoding RNA

There are numerous DNA sequences which do not code for any cellular protein, but may be expressed as non-protein coding RNA (ncRNA). These can reside between genes or within them as introns, and may take many forms. Two key forms have been indicated as playing roles in the epigenetics of pluripotency and cell identity: long noncoding RNA (lncRNA) and miRNA. While not always classed as true epigenetic factors, as they require sequence expression that is itself regulated by histone and DNA modifications, they play epigenetic-like roles in altering gene expression profiles and in some cases are integral parts of the epigenetic systems that set and maintain repressive/active patterns.

LncRNA refers to expressed, non-protein coding sequences that are longer than 200 nt. These are estimated to constitute many thousands, perhaps hundreds of thousands, of expressed sequences found in mammals (Birney et al. 2007; Kapranov et al. 2007). While they may be processed in many ways, including as precursors to miRNAs, their role in conjunction with epigenetic mechanisms is becoming clear. Two well known lncRNAs with developmental significance are X-inactive transcript (Xist), which silences X-chromosomes by coating the one from which it was transcribed (cis-acting), and HOX antisense intergenic RNA (HOTAIR) which acts to silence genes on another chromosome (trans-acting). MiRNAs originate as long RNA sequences that undergo enzymatic processing inside the nucleus (by DGCR8/Drosha) and outside (by TRBP/Dicer), leaving short (~ 22 nt) segments which are taken up by RNA-induced silencing complex (RISC) (Mallanna and Rizzino 2010). RISC then uses miRNAs to target DNA for transcriptional blocking or degradation. MiRNAs are recognized as potentially powerful regulators of gene expression and cell identity. For example, the transgene LIN28 used in iPSC protocols is an RNA binding protein that appears to exert its effect by reducing levels of the miRNA Let-7. Experimentally generated Dicer-null cells cannot create mature miRNA, and developmental abnormalities are observed (Li and Jin 2010; Mallanna and Rizzino 2010). This shows that miRNAs, as other epigenetic factors, are necessary for proper development and health.

2.1.4 Common Epigenetic Traits

Regardless of cell type, the epigenetic mark H3K36me3 is enriched across transcribed regions, following the active H3K4me3 mark (Mikkelsen et al. 2007). This makes sense given its relation to RNA polymerase II transcription, specifically elongation. Its presence was much less prominent in bivalent genes, which contain mixed active and repressive marks, and little overlap was seen between H3K36me3 and repressive H3K27me3 marks. It has been argued that these properties make H3K36me3 a useful epigenomic mapping target for expressed sequences, as "transcriptional units", particularly ncRNAs. It was also found that overlapping areas of H3K4me3 and H3K9me3 indicate imprinting regions in ESCs. Another mapping experiment, found that H3K27ac occurs in "peaks" like the active H3K4me3 mark, suggesting a similarity between active marks and resulting chromatin structure regardless of acetylation or methylation (Hawkins et al. 2010). Together these point to common epigenetic signals that could be exploited to determine current gene or chromatin status within a cell.

2.2 Epigenetic Control in De-, Re-, and Trans-Differentiation

Creation of iPSCs requires the rearrangement of existing repressive and activating marks, across the entire genome, back to those held by ESCs. The change of any mark could be viewed as a single de-differentiating event, with success affected by the stability of the mark, as well as ongoing mechanisms that work to sustain the mark. Re- and trans-differentiation constitute the same type of process as de-differentiation, being the rearrangement of required marks to adjust chromatin structure and so gene expression of a desired cell type. Therefore, epigenetic maps of iPSCs, and conclusions based from them, will be used to represent all reprogramming methods.

The most significant problem in any reprogramming method will be the issue of epigenetic accuracy. There are basically three states a gene can be in: *off*, *on*, and *poised* (Fig. 2.1). It is that third category that makes things less than straightforward, as reprogramming cannot be as simple as removing all repressive or activating marks. It is somewhat inconvenient that the most important genes for proper development need to be reset to *poised*, with an active equilibrium of repressive and activating marks (Hawkins et al. 2010). It is also possible for the process of reprogramming to set entirely new marks, unrelated to desired status. Thus while a reprogrammed cell might show adequate gene expression levels throughout most of its genome, indicating success, a bivalent gene not set properly can generate

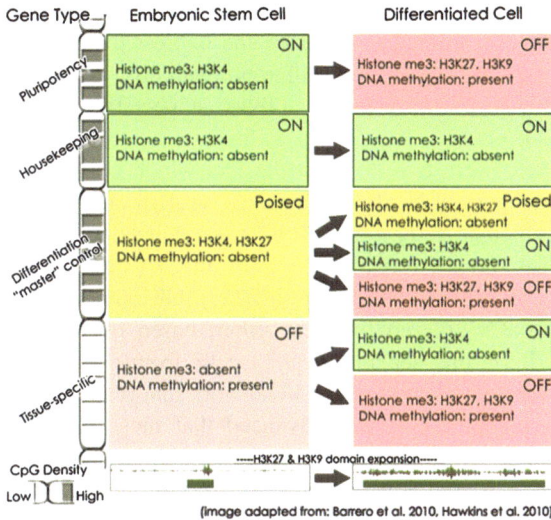

Fig. 2.1 Epigenomic map of embryonic stem and differentiated cells. Genomes are broken into regions with low/high CpG content at gene promoters, and further by gene types. Changes from pluripotent ESC state to full differentiation are described. At the *bottom*, expansion of repressive domains (ex. H3K27me3) during differentiation is shown. IPSC reprogramming must reset all marks properly

functional problems later when environmental conditions change. This is supported by a report where aberrant silencing of just one gene cluster affected proper cell function (Stadtfeld et al. 2010). Another study of ESC and iPSC methylomes showed that developmental genes of iPSCs were in a methylated state somewhere between ESCs and fibroblasts. Supporting theories that reprogramming can actively place improper epigenetic signatures, aberrant marks unrelated to ESCs or source cells were detected (Doi et al. 2009). Also, significant repressive marks naturally gained during differentiation, like H3K9me3, are removed inefficiently or even added to during de-differentiation (Mikkelsen et al. 2007). As it is, the expansion of H3K9 and H3K27me3 domains during differentiation could potentially hinder full reprogramming due to retention of some portion of the expanded domains (Hawkins et al. 2010). A recent report with detailed (single base pair resolution) epigenetic comparisons of ESCs, iPSCs, and somatic cells, showed that consistent aberrant marks were generally localized to centromeres and telomeres (Lister et al. 2011). This suggests another, in this case mechanical, problem for complete reprogramming in selected areas of chromosomes, which can affect various genes. As with other studies, Lister et al. (2011) found instances of both incomplete reprogramming and newly imposed improper epigenetic marks. Re-differentiation bias of iPSCs has also been investigated from an epigenetic standpoint (Kim et al. 2010). This revealed that DNA methylation signatures from the original cell source remained in iPSCs. These signatures appeared capable of hindering iPSC re-differentiation to other cell types, while positively affecting re-differentiation to initial cell lineage. Lister et al. (2011) re-differentiated iPSC lines and found that aberrant marks were passed on to daughter cells, which could affect cell fate decisions as well as function (though this was not proven).

MiRNAs can also be used to distinguish between cell types. 49 human cell lines were examined to compare the expression levels of 330 miRNAs (466 in some cases) across hESCs, differentiated cells, hiPSCs, and cancer cells (Neveu et al. 2010). Contrary to the authors' expectations, statistical analysis generated four clusters of cell types. In addition to healthy differentiated cells and cancer cells, there were two different clusters which both contained hESCs and hiPSCs. The grouping of hESCs and hiPSCs together suggests that reprogramming was largely sufficient to regenerate ESC miRNA levels. This is consistent with other research that showed few differences in miRNA levels between hESCs and hiPSCs (Chin et al. 2009; Stadtfeld et al. 2010). Of interest was that one cluster of hESCs and hiPSCs differed from the other, based on some of the same miRNA profiling criteria that separate healthy cells from cancer cells (Neveu et al. 2010). These were connected to the p53 network that regulates cell growth and division, as well as apoptosis. It was advanced that these two pluripotent clusters could create subdivisions of pluripotent cells, similar to naïve versus primed, based on the integrity of their p53 network. A connection was also made between the TF cMyc, which regulates cell growth and division, and iPSC clustering with cancer cells. In their analysis, down regulation of cMyc resulted in a categorical shift away from cancer, suggesting that iPSCs that clustered with cancer might have higher levels of c-Myc (Neveu et al. 2010). As c-Myc is down regulated in later stages of reprogramming, where Oct4 becomes more important for establishing pluripotency, high levels of c-Myc in iPSCs could be an indicator that they are still in an early stage and not yet fully pluripotent. Thus miRNA profiles with similarities to cancer cells could be used as a marker of incomplete reprogramming in iPSCs (and ESCs which have moved toward "primed" state, or gained cancer characteristics).

2.3 Using Epigenetics to Aid Reprogramming

Huangfu et al. (2008a) were the first to test the use of epigenetic inhibitors on induction of pluripotency. Given the repressive nature of Dnmts and HDACs, it was theorized that their inhibition might aid reprogramming by easing repression and so opening chromatin to TFs. During induction of pluripotency using four TFs (OSKM), 5′-azacytidine (5′-azaC) was used to block Dnmt processing, while suberoylanilidehydroxamide acid (SAHA), trichostatin A (TSA), and VPA were used to block HDACs. All improved efficiency to some degree, however 5′-azac and VPA proved the most useful. VPA in particular had dramatic results, improving efficiency 50–100 fold (to above 2%) while reducing required reprogramming time from 30 days down to 2 weeks and successfully replacing the TF c-Myc. The same research group later reported VPA's ability to replace both Klf4 and c-Myc, reducing needed TFs further to just Oct4 and Sox2 (Huangfu et al. 2008b). Another paper reviewing small molecules used to improve reprogramming, reported three additional inhibitors: BIX-01294 that targets G9a HMTase, RG108 targeting Dnmts, and Parnate that targets lysine-specific HDMTs (Li and Ding 2010).

2.3 Using Epigenetics to Aid Reprogramming

Improvements in reprogramming duration and efficiency due to epigenetic inhibitors were detailed in a recent review (Wang et al. 2010). Of all inhibitors examined, BIX and VPA appear to be the most powerful. Another lab tried to create a TF-free iPSC reprogramming method, which was entirely epigenetic (Han et al. 2010). The concept was to use ESC extracts, in combination with Dnmt and HDAC inhibitors. In this case 5′-azaC and TSA were used to treat somatic cells prior to transfer of extract. While full reprogramming was not accomplished, partial pluripotency was achieved, with cells apparently re-differentiating successfully. Dnmt and HDAC inhibitors were reported to improve extract performance. An earlier attempt, using only 5′-azaC and TSA on neurosphere cells, provided similar results (Ruau et al. 2008). There, histone acetylation and DNA demethylation were found to be promiscuous with respect to chromosomes, however genes associated with pluripotency were upregulated (ex. Oct4, Sox2, Klf4, Nanog, and cMyc). More interesting, while full pluripotency was not induced, neurosphere cells gained sufficient plasticity to re-differentiate to hematopoietic cells.

Given the greater efficiency ($\sim 45\%$), speed (1–2 days), and cell-division independence of SCNT and cell fusion reprogramming compared to TF-based methods (Han et al. 2010; Markoulaki et al. 2008), it is clear that ESCs and oocytes contain active epigenetic reprogramming components. One group screened extracts from pluripotent mouse ESCs and found two components, Brg1 and Baf155, associated with reactivation of Oct4 (Singhal et al. 2010). Both components were identified as subunits of the SWI/SNF complex, involved with nucleosome rearrangement as well as gene regulation via collaboration with epigenetic repressors (e. DNMT3a/b). When included as transgenes along with OSKM, or OSK, the components Brg1 and Baf155 improved induction efficiency (up to 12 fold). Their method of epigenetic action was also investigated. Euchromatin was increased, generally through enrichment of active marks such as H4Kme3 and H3K9ac. It was suggested that these subunits may have interfered with somatic SWI/SNF complexes and/or formed other complexes capable of enhancing reprogramming.

Two separate studies have recently proposed activation-induced cytidine deaminase (AID) as the active DNA demethylation component in mammals. Rather than direct demethylation of CpGs, AID would bind to 5-methylcytosine, followed by deamination of 5 mC to a thymine residue. The resulting T-G mismatch would then be repaired (i.e. return of Cytosine) by the base excision repair pathway (BER). One study used mammalian primordial germ cells (PGCs), which normally undergo DNA demethylation as part of their developmental process (Popp et al. 2010). AID deficient PGCs showed greater global methylation (3x) as compared to wild type, implicating their role in the demethylation process. It was suggested that this could be a mechanism for preventing transmission of epigenetic markings to offspring. The other study focused more specifically on reprogramming factors within ESCs. The authors took a novel approach and produced interspecies heterokaryons, specifically fusing mESCs with human fibroblasts, then knocked down AID using small interfering RNA (siRNA) (Bhutani et al. 2010).

Under normal conditions mouse ESCs were capable of inducing pluripotency in human fibroblasts quickly (1d, no celldivision), and efficiently (70%). On transient knockdown of AID, induction of pluripotency (specifically expression of Oct4 and Nanog) was greatly inhibited (80%). This was true even at 35% knockdown of AID, indicating great sensitivity to its absence, and arguing for its necessity in normal mammalian gene regulation. In support of this, CpG areas of Oct4 and Nanog promoters in knockdowns were found to retain heavy methylation. When human AID was over expressed, prior to knockdown of mouse AID, demethylation activity was rescued to some degree. Nanog showed expected demethylation of promoters and upregulated gene expression. Oct4 exhibited only partial recovery to normal levels. Further, using ChIP (chromatin immunoprecipitation), the authors found that AID binds to heavily methylated promoters (specifically Oct4 and Nanog) in fibroblasts (where DNA demethylation is required) while not at hypomethylated promoters in mESCs. Taken together, both studies paint a very strong picture of AID being one of the key natural epigenetic reprogramming mechanisms, likely required for fast and efficient iPSC methods. However it remains to be determined what the exact follow up step to deamination involves, and if there are additional players required for both activating AID or repair of T-G mismatch (Agarwal and Daley 2010; Deng 2010).

As a note of caution, another study attempting to define the demethylation components in PGCs, similar to Popp et al. (2010), came to a different conclusion about the importance of AID (Hajkova et al. 2010). As zygotes were used in place of PGCs, their results were just as pertinent to the findings of Bhutani et al. (2010). This group tracked methylation changes in the maternal and paternal pronucleus, and its relation to BER (including knockdown using inhibitors of BER). They determined that deamination was not required, and that single strand DNA breaks were sufficient to activate BER, with resulting DNA demethylation. AID might have played a role, but was dismissed based on not finding evidence for it within the zygote at that stage. Curiously the authors cite Popp et al. (2010) in support, stating that AID in that report was shown to have little effect in PGCs, but that appears inconsistent with the written conclusions of Popp et al.. Of course it is possible that AID does not play a role in PGCs or primitive zygotes, but does so in later ESC tissue. Finally, both Popp et al. and Hajkova et al. propose Tet 5mC-hydroxylases as another potential mechanism for DNA demethylation. Clearly, further research is required to sort these possibilities out, but it does appear that identification of active epigenetic mechanisms is approaching.

NcRNA might also play a role in enhancing reprogramming efficiency and speed. Since ncRNA is a vast field, and arguably not a true epigenetic mechanism, research into using it for reprogramming will not be covered in as much detail. Two reviews of using miRNAs for reprogramming cell identity cite active research as well as its potential (Mallanna and Rizzino 2010; Sun et al. 2010). The main concept outlined by both is that miRNAs may be used (or inhibited) to target specific pathways, and downstream effectors, to create desired effects while limiting unintended problems. Both name p53 as an example. While the p53 network can block full iPSC reprogramming, knocking out p53 leads to problems.

However, p53's hindrance to reprogramming appears to be limited to its activation of miRNA-145, which blocks the key TFs Oct4, Sox2, Klf4, and cMyc. Thus one could target miR-145 for inhibition, and potentially enhance reprogramming, while retaining p53's other desired effects. This indicates that abundant miRNAs expression in ESCs and/or inhibition of miRNAs in the source somatic cells, could aid reprogramming.

As mentioned in Chap. 1, miRNAs have moved beyond assisting standard TF-based reprogramming to become the method itself (Anokye-Danso et al. 2011; Miyoshi et al. 2011). The approach of Anokye-Danso et al. (2011) to fully reprogram mouse and human fibroblasts to pluripotency using expression of the miR-302/367 cluster was more efficient and faster than the common OSKM TF method. Intriguingly this study also showed the potential utility and limitation of epigenetic factors in assisting reprogramming. While VPA's strengths have been noted earlier, this group found it was only useful for reprogramming mouse fibroblasts. This was argued to be due to VPA's targeting of HDAC2 proteins. Mice have relatively higher levels of HDAC2 and so require VPA for miR-302/367 reprogramming, while humans already produce low levels of HDAC2 which are unaffected by VPA. This means that epigenetic assistance may have to be species as well as cell specific.

2.4 Epigenetics Affecting Neural Cell Fate

After reprogramming to pluripotency, epigenetics can aid differentiation to neural cell lineages. The developmental pathway from ESC to NSC to neuron or glia is the subject of much research (Juliandi et al. 2010). A genome-wide analysis of DNA methylation in NSCs, revealed that promoter regions of astrocytic genes are progressively demethylated from mid to late stages of development (Fig. 2.2) (Hatada et al. 2008). This holds true for the astrocyte gene GFAP whose promoter has been shown to exhibit hypermethylation in mESCs, only becoming hypomethylated on reaching a mature NSC state (Shimozaki et al. 2005; Takizawa et al. 2001). This is consistent with the general concepts already described, with DNA methylation blocking expression (in this case of astrocytes specific genes) by preventing the binding of transcription machinery. In addition, methylation of exons within GFAP gene sequences has been observed (Setoguchi et al. 2006). In this case, methyl-CpG-binding domain proteins (MBDs) can latch on to the methylated exon and prevent transcription, even if the promoter of the gene is unmethylated. This allows another level of gene control and so greater flexibility. Both active and passive mechanisms have been suggested for DNA demethylation, and so activation of astrocyte genes. An active method would involve removing methylated cytosines, perhaps by the base excision repair mechanism (Ma et al. 2009). Passive demethylation would come from blocking Dnmt1 maintenance activity, for example by nuclear factor 1A (NF1A) that has been reported to block Dnmt1 and allow astrocyte production (Namihira et al. 2009).

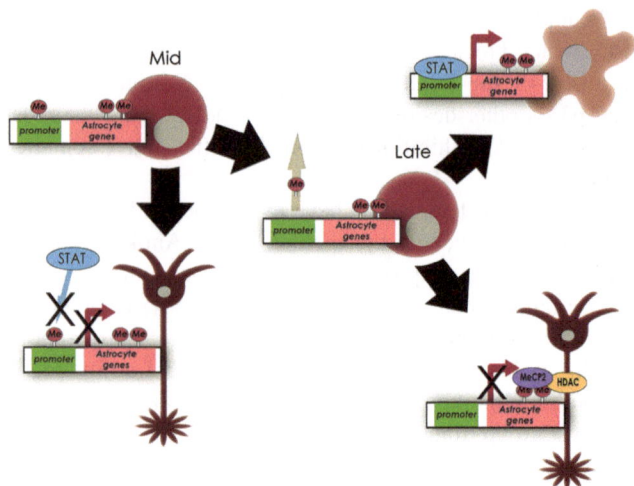

Fig. 2.2 DNA methylation regulates CNS development. Two paths of gene regulation by DNA methylation are shown. For mid-gestational NSCs, methylated promoters block transcriptional machinery preventing astrocytic gene expression. In late-gestational NSCs, methyl groups are absent from promoters but methyl CpG binding proteins may form complexes on methyl groups located within astrocyte gene sequences blocking transcription

Histone modifications play various roles during CNS development. Histone acetylation is largely associated with the opening of chromatin, and for CNS development plays a large role for neuronal genes. HDACs that remove such marks are known to prevent formation of neurons during development, and conversely HDAC inhibitors such as VPA have been shown to promote neuron production over glia in vitro (Hsieh et al. 2004). Specific HDMTs resolve bivalency, committing mESCs to NSC fates (Burgold et al. 2008). Later in development, H3K27 methylation of proneuronal genes like Neurogenin1 work to promote astrocyte production, in that particular case by inhibiting an astrocyte gene suppressor (Hirabayashi et al. 2009; Sun et al. 2001).

NcRNAs guide CNS development as well. The role of lncRNAs in regulating TFs may be particularly important for development and function of the CNS (Qureshi et al. 2010). An lncRNA known as Sox2ot was reported as being expressed in areas of active neurogenesis in adults, and that it might also regulate Sox2 TF levels during development (Amaral et al. 2009; Mercer et al. 2008). Finally, in vitro differentiation of NSCs into the oligodendritic cell lineage was improved with forced expression of lncRNA Nkx2.2AS (Tochitani and Hayashizaki 2008). Specific miRNAs are critical for early stages of neuronal lineage commitment and neural tube closure (ex. let-7), as well as regulating the proliferation and commitment of NSCs and NPCs to neural or glial fates (ex. miR-9, miR-23, and miR-125(b)) (Li and Jin 2010). MiR-124a enables RISC to degrade glia transcripts, blocking commitment to non-neuronal fates (Conaco et al. 2006). Developing cells from ESCs to NSCs express a protein (NRSF; neural-restrictive

2.4 Epigenetics Affecting Neural Cell Fate

silencer factor, also known as REST) that down regulates both neuronal genes and miR-124a, preventing premature generation of neurons. As development continues another miRNA, miR-9, may be expressed targeting NRSF. This lifts repression of neuronal genes and miR-124a, the latter suppressing non-neuronal transcripts, thereby enhancing neuronal commitment (Conaco et al. 2006; Packer et al. 2008). One target of miR-124a is the TF Sox9, which is crucial to glial lineage commitment (Cheng et al. 2009), while both miR-124a and miR-9 are capable of blocking generation of astrocytes by inhibiting STAT3 (signal transducer and activator of transcription 3) (Krichevsky et al. 2006). There are additional miRNAs, and likely more will be found, that are specific to neuronal or glial cell lineages (Juliandi et al. 2010). In fact, one study implicated the importance of a single miRNA (miR-219) in the switch from oligodendocyte precursor cell (OPC) to fully myelinating oligodendrocyte (Dugas et al. 2010).

Liu et al. (2010) studied trans-differentiation of fibroblasts to myelinating oligodendrocytes by using ncRNAs in conjunction with true epigenetic reprogramming mechanisms. While they failed to produce myelinating oligodendrocytes, they did find that Dmnt and HDAC inhibitors (5′-azaC and TSA, respectively) were able to induce endogenous expression of oligodendrocyte-related genes without use of TFs (Liu et al. 2010; Sun et al. 2010). One potential roadblock to trans-differentiation was the unintended de-repression of transcriptional inhibitors of myelin genes by 5′-azaC and TSA. The authors suggested using miRNAs targeting the myelin inhibiting genes to improve trans-differentiation. Whether a combination of miRNAs, 5′-azaC, and TSA alone would be enough to help trans-differentiation into myelinating oligodendrocytes needs to be tested. Another possibility suggested involves adding SWI/SNF subunits or other chromatin remodelling complexes during reprogramming.

Similarly, and more effectively, Ambasudhan et al. (2011) used miRNA-124 in combination with two TFs (MYT1L and BRYN2) to generate functional neurons from adult human fibroblasts (Ambasudhan et al. 2011). This builds on the previously discussed finding that miRNA-124 aids in the development of neurons by hindering non-neuronal fate decisions. For trans-differentiation to become the primary method for generating all neural cell lineages, a better understanding of how epigenetic mechanisms regulate cell identity is required.

2.5 A Caveat

The size of mammalian genomes creates a practical difficulty in getting both exact and comprehensive data. Some researchers have found it easier, and more productive, to select regions of the genome where one expects to see important results. For histone modifications, being selective becomes especially important, given the vast number of possibilities. However, this will affect the quality of results, as well as ability to compare results. Further, when analyzing large scale and genome-wide datasets, important details can be lost, or unimportant details magnified, depending on statistical methods. This was seen in an earlier study when statistical methods created conflicting results (Chin et al. 2009; Loh and Lim 2010). Clear and uniform statistical methods should be derived for epigenomic research to reduce potential conflicts. Finally, results by Meissner et al. (2008) indicate methylation level increases during cell culturing, causing potential aberrant hypermethylation. They argue that this may affect results from in vitro cell models. All of these should be considered when evaluating epigenetic maps.

References

Agarwal S, Daley GQ (2010) AID for reprogramming. Cell Res 20:253–255

Amaral PP, Neyt C, Wilkins SJ, skarian-Amiri ME, Sunkin SM, Perkins AC, Mattick JS (2009) Complex architecture and regulated expression of the Sox2ot locus during vertebrate development. RNA 15:2013–2027

Ambasudhan R, Talantova M, Coleman R, Yuan X, Zhu S, Lipton SA, Ding S (2011) Direct reprogramming of adult human fibroblasts to functional neurons under defined conditions. Cell Stem Cell 9:113–118

Anokye-Danso F, Trivedi CM, Juhr D, Gupta M, Cui Z, Tian Y, Zhang Y, Yang W, Gruber PJ, Epstein JA, Morrisey EE (2011) Highly efficient miRNA-mediated reprogramming of mouse and human somatic cells to pluripotency. Cell Stem Cell 8:376–388

Bhutani N, Brady JJ, Damian M, Sacco A, Corbel SY, Blau HM (2010) Reprogramming towards pluripotency requires AID-dependent DNA demethylation. Nature 463:1042–1047

Birney E, Stamatoyannopoulos JA, Dutta A, Guigo R, Gingeras TR, Margulies EH, Weng Z, Snyder M, Dermitzakis ET, Thurman RE, Kuehn MS, Taylor CM, Neph S, Koch CM, Asthana S, Malhotra A, Adzhubei I, Greenbaum JA, Andrews RM, Flicek P, Boyle PJ, Cao H, Carter NP, Clelland GK, Davis S, Day N, Dhami P, Dillon SC, Dorschner MO, Fiegler H, Giresi PG, Goldy J, Hawrylycz M, Haydock A, Humbert R, James KD, Johnson BE, Johnson EM, Frum TT, Rosenzweig ER, Karnani N, Lee K, Lefebvre GC, Navas PA, Neri F, Parker SC, Sabo PJ, Sandstrom R, Shafer A, Vetrie D, Weaver M, Wilcox S, Yu M, Collins FS, Dekker J, Lieb JD, Tullius TD, Crawford GE, Sunyaev S, Noble WS, Dunham I, Denoeud F, Reymond A, Kapranov P, Rozowsky J, Zheng D, Castelo R, Frankish A, Harrow J, Ghosh S, Sandelin A, Hofacker IL, Baertsch R, Keefe D, Dike S, Cheng J, Hirsch HA, Sekinger EA, Lagarde J, Abril JF, Shahab A, Flamm C, Fried C, Hackermuller J, Hertel J, Lindemeyer M, Missal K, Tanzer A, Washietl S, Korbel J, Emanuelsson O, Pedersen JS, Holroyd N, Taylor R, Swarbreck D, Matthews N, Dickson MC, Thomas DJ, Weirauch MT, Gilbert J, Drenkow J, Bell I, Zhao X, Srinivasan KG, Sung WK, Ooi HS, Chiu KP, Foissac S, Alioto T, Brent M, Pachter L, Tress ML, Valencia A, Choo SW, Choo CY, Ucla C, Manzano C, Wyss C, Cheung E, Clark TG, Brown JB, Ganesh M, Patel S, Tammana H, Chrast J, Henrichsen CN, Kai C, Kawai J, Nagalakshmi U, Wu J, Lian Z, Lian J, Newburger P, Zhang X, Bickel P, Mattick JS, Carninci P, Hayashizaki Y, Weissman S, Hubbard T, Myers RM, Rogers J, Stadler PF, Lowe TM, Wei CL, Ruan Y, Struhl K, Gerstein M, Antonarakis SE, Fu Y, Green ED, Karaoz U, Siepel A, Taylor J, Liefer LA, Wetterstrand KA, Good PJ, Feingold EA, Guyer MS, Cooper GM, Asimenos G, Dewey CN, Hou M, Nikolaev S, Montoya-Burgos JI, Loytynoja A, Whelan S, Pardi F, Massingham T, Huang H, Zhang NR, Holmes I, Mullikin JC, Ureta-Vidal A, Paten B, Seringhaus M, Church D, Rosenbloom K, Kent WJ, Stone EA, Batzoglou S, Goldman N, Hardison RC, Haussler D, Miller W, Sidow A, Trinklein ND, Zhang ZD, Barrera L, Stuart R, King DC, Ameur A, Enroth S, Bieda MC, Kim J, Bhinge AA, Jiang N, Liu J, Yao F, Vega VB, Lee CW, Ng P, Shahab A, Yang A, Moqtaderi Z, Zhu Z, Xu X, Squazzo S, Oberley MJ, Inman D, Singer MA, Richmond TA, Munn KJ, Rada-Iglesias A, Wallerman O, Komorowski

J, Fowler JC, Couttet P, Bruce AW, Dovey OM, Ellis PD, Langford CF, Nix DA, Euskirchen G, Hartman S, Urban AE, Kraus P, Van CS, Heintzman N, Kim TH, Wang K, Qu C, Hon G, Luna R, Glass CK, Rosenfeld MG, Aldred SF, Cooper SJ, Halees A, Lin JM, Shulha HP, Zhang X, Xu M, Haidar JN, Yu Y, Ruan Y, Iyer VR, Green RD, Wadelius C, Farnham PJ, Ren B, Harte RA, Hinrichs AS, Trumbower H, Clawson H (2007) Identification and analysis of functional elements in 1% of the human genome by the ENCODE pilot project. Nature 447:799–816

Burgold T, Spreafico F, De SF, Totaro MG, Prosperini E, Natoli G, Testa G (2008) The histone H3 lysine 27-specific demethylase Jmjd3 is required for neural commitment. PLoS One 3:e3034

Cheng LC, Pastrana E, Tavazoie M, Doetsch F (2009) miR-124 regulates adult neurogenesis in the subventricular zone stem cell niche. Nat Neurosci 12:399–408

Chin MH, Mason MJ, Xie W, Volinia S, Singer M, Peterson C, Ambartsumyan G, Aimiuwu O, Richter L, Zhang J, Khvorostov I, Ott V, Grunstein M, Lavon N, Benvenisty N, Croce CM, Clark AT, Baxter T, Pyle AD, Teitell MA, Pelegrini M, Plath K, Lowry WE (2009) Induced pluripotent stem cells and embryonic stem cells are distinguished by gene expression signatures. Cell Stem Cell 5:111–123

Conaco C, Otto S, Han JJ, Mandel G (2006) Reciprocal actions of REST and a microRNA promote neuronal identity. Proc Natl Acad Sci U S A 103:2422–2427

Deng W (2010) AID in reprogramming: quick and efficient: identification of a key enzyme called AID, and its activity in DNA demethylation, may help to overcome a pivotal epigenetic barrier in reprogramming somatic cells toward pluripotency. Bioessays 32:385–387

Doi A, Park IH, Wen B, Murakami P, Aryee MJ, Irizarry R, Herb B, Ladd-Acosta C, Rho J, Loewer S, Miller J, Schlaeger T, Daley GQ, Feinberg AP (2009) Differential methylation of tissue- and cancer-specific CpG island shores distinguishes human induced pluripotent stem cells, embryonic stem cells and fibroblasts. Nat Genet 41:1350–1353

Dugas JC, Cuellar TL, Scholze A, Ason B, Ibrahim A, Emery B, Zamanian JL, Foo LC, McManus MT, Barres BA (2010) Dicer1 and miR-219 are required for normal oligodendrocyte differentiation and myelination. Neuron 65:597–611

Hajkova P, Jeffries SJ, Lee C, Miller N, Jackson SP, Surani MA (2010) Genome-wide reprogramming in the mouse germ line entails the base excision repair pathway. Science 329:78–82

Han J, Sachdev PS, Sidhu KS (2010) A combined epigenetic and non-genetic approach for reprogramming human somatic cells. PLoS One 5:e12297

Hatada I, Namihira M, Morita S, Kimura M, Horii T, Nakashima K (2008) Astrocyte-specific genes are generally demethylated in neural precursor cells prior to astrocytic differentiation. PLoS One 3:e3189

Hawkins RD, Hon GC, Lee LK, Ngo Q, Lister R, Pelizzola M, Edsall LE, Kuan S, Luu Y, Klugman S, Antosiewicz-Bourget J, Ye Z, Espinoza C, Agarwahl S, Shen L, Ruotti V, Wang W, Stewart R, Thomson JA, Ecker JR, Ren B (2010) Distinct epigenomic landscapes of pluripotent and lineage-committed human cells. Cell Stem Cell 6:479–491

Hirabayashi Y, Suzki N, Tsuboi M, Endo TA, Toyoda T, Shinga J, Koseki H, Vidal M, Gotoh Y (2009) Polycomb limits the neurogenic competence of neural precursor cells to promote astrogenic fate transition. Neuron 63:600–613

Hsieh J, Nakashima K, Kuwabara T, Mejia E, Gage FH (2004) Histone deacetylase inhibition-mediated neuronal differentiation of multipotent adult neural progenitor cells. Proc Natl Acad Sci U S A 101:16659–16664

Huangfu D, Maehr R, Guo W, Eijkelenboom A, Snitow M, Chen AE, Melton DA (2008a) Induction of pluripotent stem cells by defined factors is greatly improved by small-molecule compounds. Nat Biotechnol 26:795–797

Huangfu D, Osafune K, Maehr R, Guo W, Eijkelenboom A, Chen S, Muhlestein W, Melton DA (2008b) Induction of pluripotent stem cells from primary human fibroblasts with only Oct4 and Sox2. Nat Biotechnol 26:1269–1275

Jackson M, Krassowska A, Gilbert N, Chevassut T, Forrester L, Ansell J, Ramsahoye B (2004) Severe global DNA hypomethylation blocks differentiation and induces histone hyperacetylation in embryonic stem cells. Mol Cell Biol 24:8862–8871

Juliandi B, Abematsu M, Nakashima K (2010) Epigenetic regulation in neural stem cell differentiation. Dev Growth Differ 52:493–504

Kapranov P, Cheng J, Dike S, Nix DA, Duttagupta R, Willingham AT, Stadler PF, Hertel J, Hackermuller J, Hofacker IL, Bell I, Cheung E, Drenkow J, Dumais E, Patel S, Helt G, Ganesh M, Ghosh S, Piccolboni A, Sementchenko V, Tammana H, Gingeras TR (2007) RNA maps reveal new RNA classes and a possible function for pervasive transcription. Science 316:1484–1488

Kim K, Doi A, Wen B, Ng K, Zhao R, Cahan P, Kim J, Aryee MJ, Ji H, Ehrlich LI, Yabuuchi A, Takeuchi A, Cunniff KC, Hongguang H, Kinney-Freeman S, Naveiras O, Yoon TJ, Irizarry RA, Jung N, Seita J, Hanna J, Murakami P, Jaenisch R, Weissleder R, Orkin SH, Weissman IL, Feinberg AP, Daley GQ (2010) Epigenetic memory in induced pluripotent stem cells. Nature 467:285–290

Krichevsky AM, Sonntag KC, Isacson O, Kosik KS (2006) Specific microRNAs modulate embryonic stem cell-derived neurogenesis. Stem Cells 24:857–864

Li W, Ding S (2010) Small molecules that modulate embryonic stem cell fate and somatic cell reprogramming. Trends Pharmacol Sci 31:36–45

Li X, Jin P (2010) Roles of small regulatory RNAs in determining neuronal identity. Nat Rev Neurosci 11:329–338

Lister R, Pelizzola M, Kida YS, Hawkins RD, Nery JR, Hon G, ntosiewicz-Bourget J, O'Malley R, Castanon R, Klugman S, Downes M, Yu R, Stewart R, Ren B, Thomson JA, Evans RM, Ecker JR (2011) Hotspots of aberrant epigenomic reprogramming in human induced pluripotent stem cells. Nature 471:68–73

Liu J, Sandoval J, Doh ST, Cai L, Lopez-Rodas G, Casaccia P (2010) Epigenetic modifiers are necessary but not sufficient for reprogramming non-myelinating cells into myelin gene-expressing cells. PLoS One 5:e13023

Loh KM, Lim B (2010) Recreating pluripotency? Cell Stem Cell 7:137–139

Ma DK, Guo JU, Ming GL, Song H (2009) DNA excision repair proteins and Gadd45 as molecular players for active DNA demethylation. Cell Cycle 8:1526–1531

Mallanna SK, Rizzino A (2010) Emerging roles of microRNAs in the control of embryonic stem cells and the generation of induced pluripotent stem cells. Dev Biol 344:16–25

Markoulaki S, Meissner A, Jaenisch R (2008) Somatic cell nuclear transfer and derivation of embryonic stem cells in the mouse. Methods 45:101–114

Meissner A, Mikkelsen TS, Gu H, Wernig M, Hanna J, Sivachenko A, Zhang X, Bernstein BE, Nusbaum C, Jaffe DB, Gnirke A, Jaenisch R, Lander ES (2008) Genome-scale DNA methylation maps of pluripotent and differentiated cells. Nature 454:766–770

Mercer TR, Dinger ME, Sunkin SM, Mehler MF, Mattick JS (2008) Specific expression of long noncoding RNAs in the mouse brain. Proc Natl Acad Sci U S A 105:716–721

Mikkelsen TS, Ku M, Jaffe DB, Issac B, Lieberman E, Giannoukos G, Alvarez P, Brockman W, Kim TK, Koche RP, Lee W, Mendenhall E, O'Donovan A, Presser A, Russ C, Xie X, Meissner A, Wernig M, Jaenisch R, Nusbaum C, Lander ES, Bernstein BE (2007) Genome-wide maps of chromatin state in pluripotent and lineage-committed cells. Nature 448:553–560

Miyoshi N, Ishii H, Nagano H, Haraguchi N, Dewi DL, Kano Y, Nishikawa S, Tanemura M, Mimori K, Tanaka F, Saito T, Nishimura J, Takemasa I, Mizushima T, Ikeda M, Yamamoto H, Sekimoto M, Doki Y, Mori M (2011) Reprogramming of mouse and human cells to pluripotency using mature microRNAs. Cell Stem Cell 8:633–638

Namihira M, Kohyama J, Semi K, Sanosaka T, Deneen B, Taga T, Nakashima K (2009) Committed neuronal precursors confer astrocytic potential on residual neural precursor cells. Dev Cell 16:245–255

Neveu P, Kye MJ, Qi S, Buchholz DE, Clegg DO, Sahin M, Park IH, Kim KS, Daley GQ, Kornblum HI, Shraiman BI, Kosik KS (2010) MicroRNA profiling reveals two distinct p53-related human pluripotent stem cell states. Cell Stem Cell 7:671–681

Packer AN, Xing Y, Harper SQ, Jones L, Davidson BL (2008) The bifunctional microRNA miR-9/miR-9* regulates REST and CoREST and is downregulated in Huntington's disease. J Neurosci 28:14341–14346

Popp C, Dean W, Feng S, Cokus SJ, Andrews S, Pellegrini M, Jacobsen SE, Reik W (2010) Genome-wide erasure of DNA methylation in mouse primordial germ cells is affected by AID deficiency. Nature 463:1101–1105

Qureshi IA, Mattick JS, Mehler MF (2010) Long non-coding RNAs in nervous system function and disease. Brain Res 1338:20–35

Ruau D, Ensenat-Waser R, Dinger TC, Vallabhapurapu DS, Rolletschek A, Hacker C, Hieronymus T, Wobus AM, Muller AM, Zenke M (2008) Pluripotency associated genes are reactivated by chromatin-modifying agents in neurosphere cells. Stem Cells 26:920–926

Setoguchi H, Namihira M, Kohyama J, Asano H, Sanosaka T, Nakashima K (2006) Methyl-CpG binding proteins are involved in restricting differentiation plasticity in neurons. J Neurosci Res 84:969–979

Shimozaki K, Namihira M, Nakashima K, Taga T (2005) Stage- and site-specific DNA demethylation during neural cell development from embryonic stem cells. J Neurochem 93:432–439

Singhal N, Graumann J, Wu G, rauzo-Bravo MJ, Han DW, Greber B, Gentile L, Mann M, Scholer HR (2010) Chromatin-remodeling components of the BAF complex facilitate reprogramming. Cell 141:943–955

Stadtfeld M, Apostolou E, Akutsu H, Fukuda A, Follett P, Natesan S, Kono T, Shioda T, Hochedlinger K (2010) Aberrant silencing of imprinted genes on chromosome 12qF1 in mouse induced pluripotent stem cells. Nature 465:175–181

Sun X, Fu X, Han W, Zhao Y, Liu H (2010) Can controlled cellular reprogramming be achieved using microRNAs? Ageing Res Rev 9:475–483

Sun Y, Nadal-Vicens M, Misono S, Lin MZ, Zubiaga A, Hua X, Fan G, Greenberg ME (2001) Neurogenin promotes neurogenesis and inhibits glial differentiation by independent mechanisms. Cell 104:365–376

Surface LE, Thornton SR, Boyer LA (2010) Polycomb group proteins set the stage for early lineage commitment. Cell Stem Cell 7:288–298

Takizawa T, Nakashima K, Namihira M, Ochiai W, Uemura A, Yanagisawa M, Fujita N, Nakao M, Taga T (2001) DNA methylation is a critical cell-intrinsic determinant of astrocyte differentiation in the fetal brain. Dev Cell 1:749–758

Tochitani S, Hayashizaki Y (2008) Nkx2.2 antisense RNA overexpression enhanced oligodendrocytic differentiation. Biochem Biophys Res Commun 372:691–696

Wang Y, Mah N, Prigione A, Wolfrum K, ndrade-Navarro MA, Adjaye J (2010) A transcriptional roadmap to the induction of pluripotency in somatic cells. Stem Cell Rev 6:282–296

Chapter 3
Prospects for Cell Replacement Therapies for Neurodegenerative Diseases

Abstract Our extensive knowledge of pluripotent SCs has been used by many researchers to develop cell replacement therapies for different diseases, including those affecting the brain. Different studies with animal models of neurological disease showed that cell grafts are able to improve clinical symptoms. The recent advances in the iPSC field encourage more neuroscientists to improve transplantation strategies for clinical purposes. Here we review preclinical studies involving cell-based therapies for Parkinson's disease (PD), and discuss the future prospects for cell replacement therapies for PD and the childhood white matter disorder, Vanishing White Matter (VWM), which disorders are our research focus.

Keywords Parkinson's disease · Vanishing white matter disease · Cell replacement therapy · ESC · iPSC

3.1 Cell Therapies for Neurodegenerative Diseases

The application of cell replacement therapy for neurodegenerative diseases including PD, Alzheimer's Disease and Amyotrophic Lateral Sclerosis, using SCs or SC-derived cells has attracted much attention (Kim and de Vellis 2009; Papadeas and Maragakis 2009; Wijeyekoon and Barker 2009). Because of their highly proliferative character and their potential to differentiate into all cell types, ESCs have long been seen as the potential cell source to repair the damaged brain (Koch et al. 2009). The introduction of iPSCs brought new excitement. The principle of iPSC based therapy can be seen in Fig. 3.1. The use of iPSCs from the patients' own somatic cells could bypass immune rejection issues and ethical concerns around the use of ESCs. However, as discussed in Chap. 1, many issues need to be addressed before iPSCs can be used as a cell source for cell therapies.

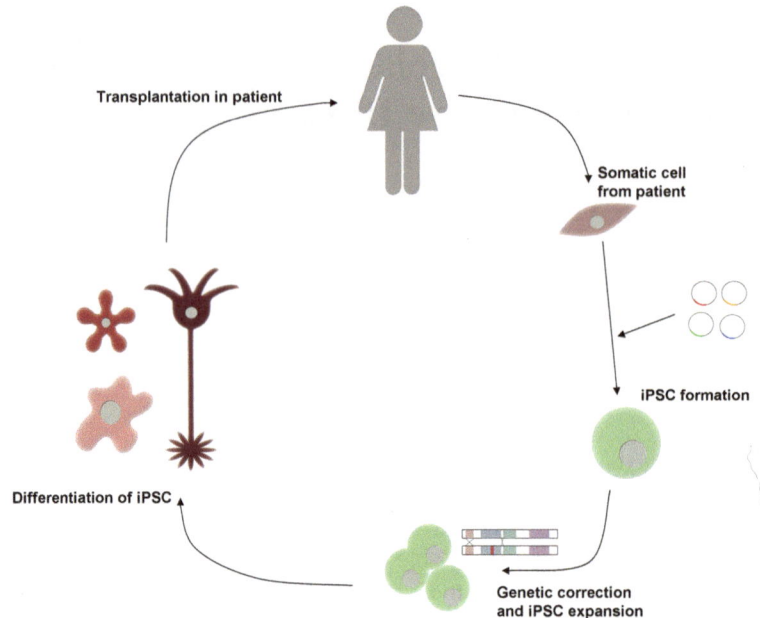

Fig. 3.1 Cell therapy with iPSC. The process of producing patient-specific cells for transplantation. First, somatic cells are derived from the patient. These cells are reprogrammed by for example viral transduction. iPSCs form and, if necessary, the mutation can be corrected by homologous recombination. iPSCs are expanded to a sufficient amount and consequently differentiated into the cell type of interest. These cells can then be transplanted into the patient

In order for cell replacement therapies to have realistic clinical prospects, ESC/iPSC-derived grafts should give lasting functional and behavioural improvement and cause no deleterious side effects in experimental animal models. Therefore we need proof-of-principle studies to show long-term cell survival and functional integration into the host brain replacing the lost cell structures. Additionally, we need to improve differentiation protocols to produce high enough cell numbers of transplantable cells. Since the different neurodegenerative diseases are characterized by different underlying mechanisms, these also have to be taken into consideration in the design of a transplantation strategy for a specific disorder. This chapter will discuss PD and VWM, current treatments and the potential of developing cell replacement therapies for these neurodegenerative diseases.

3.2 Parkinson's Disease

3.2.1 Etiology and Pathogenesis

PD is the second most common neurodegenerative disorder with an overall prevalence of approximately 2.3% in an age group ranging from 65 to 89 years in

Europe (de Rijk et al. 1997). It is also the most common movement disorder and is characterized by a progressive deterioration of motor skills, speech and cognitive functions. Tremor, rigidity and stiffness of the muscles are the most characteristic signs of people with PD (Gelb et al. 1999). Although motor symptoms are the most prominent features of PD, non-motor signs are also present. These include signs like insomnia, depression, constipation and dementia (Chaudhuri et al. 2006).

The most characteristic underlying pathology of PD is a nigrostriatal loss of dopaminergic neurons, which in turn affects the function of the basal ganglia. Ultimately, this leads to an increase in inhibition of the thalamus and subsequently of the motor cortex, causing the impairment of movement (Albin et al. 1995). The loss of dopaminergic neurons might be related to the accumulation of the protein alpha-synuclein (Maries et al. 2003), which has been found in Lewy bodies in brains of patients (Spillantini et al. 1997), known as a pathological hallmark of PD (Gibb and Lees 1989). Despite the focus on the striatum, there are also other neuronal systems affected in PD (Braak and Braak 2000), giving rise to the full clinical picture of the disease.

With our aging population, this health problem will only become more serious. Therefore, it is of great importance to develop new therapeutic approaches.

3.2.2 Treatment Options for Parkinson's Disease

Currently, there are several therapeutic approaches for PD, ranging from medication to surgical procedures. The focus of research has been the replacement of dopamine levels in the brain, which are adversely affected in PD. To begin with, the most approved and standard treatment is the administration of levodopa (L-dopa). L-dopa is a dopamine precursor and processed to dopamine in the brain. This drug has proven to dramatically improve the motor signs of PD patients. Unfortunately, with long-term use different motor complications arise (Luquin et al. 1992). Patients begin to experience involuntary movement (dyskinesia) or de-sentitize. De-sensitization is a result of decreasing L-dopa levels in the body and involves a reappearance of signs of PD. When motor complications arise, L-dopa treatment is adjusted, but finding the right doses is a serious problem for patients that experience both dyskinesia alternated with severe motor disability. Non-motor symptoms are not controlled by the drug and the degeneration still progresses and cannot be ameliorated or halted by L-dopa administration. Therefore, L-dopa does not offer a final solution or cure for PD.

Besides medication with L-dopa, other various dopamine agonists have been tested and employed as anti-PD drugs, with the hope of overcoming the undesirable dyskinesia. Montastruc et al. (1997) suggest a beneficial effect of dopamine agonists on the risk of dyskinesia. However, the other issues mentioned, such as progression of disease and the control of other complaints, remain to be solved. Other therapies developed include surgically lesioning the globus pallidus interna to interfere with the abnormal communication between the basal ganglia and the cortex, and deep-brain stimulation. These techniques are much more invasive and therefore have more risks, such as haemorrhages and infection. Others have investigated gene therapy as an alternative approach to current treatments (reviewed by Witt and Marks 2011).

3.2.3 Fetal Tissue Transplantation

Tissue replacement therapies are focused on restoring the dopaminergic signalling between the substantia nigra and the striatum. Although tissue replacement will not work against non-motor signs, deterioration of motor skills are generally viewed as the most disturbing signs.

Research has focused on directly replacing the affected dopaminergic neurons. Previous research has successfully used fetal nigral transplantation in rodents and primates (Bankiewicz et al. 1990; Brundin et al. 1986). Clinical studies reported a beneficial effect on motor signs, but only in open label trials and not in double-blind placebo-controlled designs (Kordower et al. 1995; Lindvall et al. 1992; Olanow et al. 2003). Olanow et al. (2003) observed severe dyskinesias in many patients and the overall improvement was disappointingly small. Many explanations have been proposed for the difference in outcome between these two different types of trials, ranging from a placebo effect and experimenter bias to more methodological issues, such as the development of an immune response, the difference in preparation of the graft tissue, patient selection, L-dopa responsiveness and graft placement (Hwang et al. 2010; Winkler et al. 2005).

Problems observed with replacement therapy are graft induced dyskinesias (GID). These can develop independent of the previous L-dopa induced dyskinesias and cannot be predicted from the daily dopaminergic medication level or the severeness of L-dopa induced dyskinesias (Wijesekera and Leigh 2009). One of the causes of these GIDs could be serotonergic neurons that are present in the cell grafts. Serotonergic neurons have the ability to release dopamine, but lack dopamine transporters, leaving an excess of dopamine in the synapse. Indeed, studies showed that grafts with a small amount of dopaminergic neurons relative to serotonergic neurons worsened dyskinesias (Hedlund and Perlmann 2009). The number of dopamine innervations that are left before the transplantation appear to influence the induction of dyskinesias through serotonergic neurons. If the innervations are below 20%, the serotonergic neurons are more likely to induce dyskinesias (Hedlund and Perlmann 2009). The marker used for dopaminergic differentiation, TH, is also present in serotonergic neurons, making separation of these cell types more difficult.

Fetal dopaminergic neurons survive for a long time after transplantation, but some neurons in transplants older than 10 years show α-synuclein pathology (Hedlund and Perlmann 2009). It is not clear if this represents normal ageing, a reaction on the accumulation of proteins in the cell environment or actual pathological processes in the cell itself. Inclusion bodies may function as a method to prevent toxicity from α-synuclein oligomers, and therefore represent a clearance mechanism instead of a pathological process. However, some studies showed that transplanted cells do show some pathological characteristics, e.g. a down regulation of the dopamine transporter and TH (Hedlund and Perlmann 2009). The pathology in the transplanted neurons could be caused by reactive microglia in the substantia nigra, which can cause neuroinflammation in these neurons. The significance of this pathology is not clear. It is only present in a small proportion of the transplanted cells, and it is not known if it influences the motor functions of the patients. More research is needed to identify the exact influence of a PD brain on healthy dopaminergic neurons.

3.2.4 Cell Therapy

Although the results obtained from fetal mesencephalic cell grafts differ, the positive outcomes of several studies do support the ongoing effort to develop an effective cell therapy for PD. Limited availability of fetal cells (Morizane et al. 2008) is however an issue. Therefore the use of ESCs or iPSCs in cell therapies for PD patients has been explored.

To generate a high number of functional dopaminergic cells, different labs tried to optimize in vitro protocols. Lee et al. (2000) showed that half of the murine neurons, which were obtained after ESC differentiation, adopted a ventral mid- and hindbrain fate, revealing a great efficiency. Kawasaki et al. (2002) reported a frequency of approximately 35% for generating TH-positive neurons from primate ESCs, which also released dopamine upon depolarization, suggesting that the neurons contained a considerable number of functional dopaminergic cells. A recent study by Cooper et al. (2010) reported a method of generating dopaminergic neurons with a ventral midbrain identity from both ESCs and iPSCs without the use of mouse feeder cells. Yoshizaki et al. (2004) found that among the dopaminergic neurons in their cultures, there were also various unidentified cells. They tried to separate the dopaminergic cells from these unidentified cells, by fluorescence-activated cell sorting the cells expressing a green fluorescent protein (GFP) reporter gene driven by the TH-gene promoter. When they transplanted the GFP-expressing cells into the mouse brain it led to a reduction in Parkinsonian signs on the behavioural level. However, the TH-gene reporter is also present in serotonergic cells and therefore it is likely that non-dopaminergic neurons were grafted as well. The high portion (21–65%) of SCs developing into *non*-dopaminergic cells raised questions about its usefulness for transplantation (Freed 2002). Nevertheless, the fact that SCs can differentiate into the cell type needed once they are grafted is promising.

ESCs or iPSCs are not used in clinical trials for PD yet, but animal studies have shown the beneficial effects of this surgical approach in models of PD. It was shown that mouse ESC-derived dopaminergic progenitors can improve motor function in a PD rat model (Preynat-Seauve et al. 2009). Moreover, it was demonstrated that the transplantation of human NSCs led to neuroprotection in a rat model of PD through secretion of trophic factors (Yasuhara et al. 2006). Others transplanted dopaminergic neurons obtained from monkey ESCs in a primate model of PD (Takagi et al. 2005). The monkeys were treated with 1-methyl-4-phenyl-1,2,3,6-tetrahydropyridine (MPTP) to induce PD and a beneficial effect on a behavioural level and a neurological level through positron emission tomography (PET) was observed after transplantation. However, some amelioration in behavioural signs was also observed in sham-operated primates.

Some studies performed transplantations with iPSC-derived dopaminergic cells. Wernig et al. (2008) showed that mouse iPSC-derived dopaminergic neurons improved motor function in a rat model of PD. To test whether the grafted neurons were functionally integrated in the brain, electrophysiological measurements and morphology analyses were conducted and confirmed that the grafted cells had functional, active properties. The transplantation led to an improvement in Parkinsonian signs, both functionally and behaviourally. iPSCs from PD patients using a reprogramming method based on excisable viral vectors, were successfully differentiated into dopaminergic neurons in vivo (Soldner et al. 2009). Human derived iPSCs have been differentiated into dopaminergic neurons and transplanted into the brains of Parkinsonian rats (Hargus et al. 2010; Rhee et al. 2011). Results showed that these cells are capable of reducing the motor signs, although the survival rate of the transplanted cells is often low. This low survival rate is still an unresolved issue. Takagi et al. (2005) showed the highest survival rate (1.3–2.7%), while others reported no or only few surviving dopaminergic cells in the grafted PD rodents at all (Park et al. 2005).

3.3 Childhood Brain White Matter Disorders

3.3.1 Etiology and Pathogenesis

Vanishing white matter is a progressive disease in which the white matter of the brain becomes increasingly abnormal and eventually literally disappears. The onset is generally in childhood, although there are patients with an adolescent or adult onset. Early onset of the disease is related to more rapid disease progression (van der Knaap et al. 2003). Patients with onset in the first years of life usually die within a few years. The latest known disease onset is over 60 years (Labauge et al. 2009). The disease is always fatal. VWM patients generally present with motor signs as ataxia and spasticity. Trauma to the head, fever or acute fright may motor signs such as accelerate the disease and can lead to seizures, coma and death (van der Knaap et al. 2006). Although patients can come out of their coma their recovery is often incomplete (Fig. 3.2).

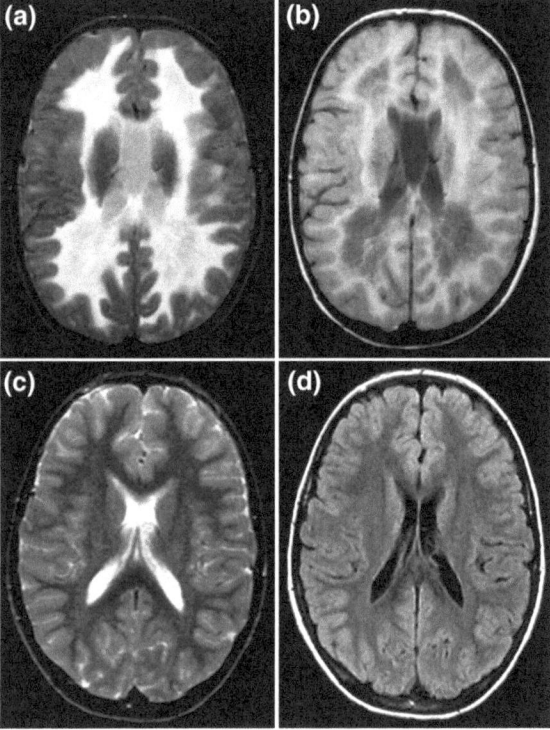

Fig. 3.2 VWM patients show an abnormal MRI pattern. This VWM patient shows a diffusely abnormal white matter (**a** T2-weighted image), which has vanished and has been replaced by tissue water (**b** FLAIR image). **c** and **d** show the comparable T2-weighted and FLAIR images of a healthy child. CSF is white on T2-weighted images; one cannot differentiate between abnormal white matter and tissue water or CSF. On FLAIR images, CSF is dark and abnormal white matter is white, making it possible to distinguish the two

On magnetic resonance imaging (MRI), patients show a diffuse abnormality of cerebral white matter, with some atrophy in the cerebellum and brainstem in later stages (van der Knaap et al. 2006). The cerebral white matter is first abnormal, then becomes rarefied and progressively cystic, and in late stages of the disease patients may have lost basically all cerebral white matter. Histopathological analyses show that the white matter has a gelatinous to cavitary appearance (van der Knaap et al. 2006). Axons are lost in the areas of cavitation, but tend to be spared in the non-cystic areas. In cavitated areas there is a significant loss of oligodendrocytes, while there is an increase in better preserved areas. The oligodendrocytes appear foamy. They fail to form a sufficient amount of myelin. Astrocytes are dysmorphic with blunt processes and fail to react properly upon the myelin and axonal loss as shown by lack of adequate gliosis (Eichler and Van Haren 2007). Both oligodendrocytes and astrocytes have an immature phenotype (Bugiani et al. 2011).

VWM patients have a mutation in any of the five genes encoding the subunits of the translational initiation factor eIF2B (*EIF2B1-5*). eIF2B activates eukaryotic initiation factor 2 (eIF2) by exchanging GDP for GTP. eIF2 is important for the initiation of translation of mRNA into peptide. During stress conditions mRNA translation is inhibited to limit the accumulation of proteins and to save energy (van der Knaap et al. 2006). This inhibition takes place through phosphorylation of eIF2, which in turn inhibits eIF2B, thereby preventing other eIF2 complexes getting activated and initiating mRNA translation. VWM patients show a lowered eIF2B activity, which might explain the episodes of deterioration after cellular stress. However, it is not clear why glial cells are predominantly affected by the disease.

3.3.2 Prospect for Cell Therapy

Currently, there is no treatment available for VWM. Because of the widespread dysfunction and loss of glial cells, especially in view of the much better preservation of neurons and axons, cell replacement therapy is viewed as a promising option for these patients.

The cerebral white matter of children with VWM is diffusely affected. Because such a large part of the brain is affected, cell grafts need to involve a large number of cells. Fetal tissue is only scarcely available. The use of ESC- or iPSC-derived cells has therefore better prospects, as these can be expanded in vitro before transplantation. Since the disease-causing mutations are known, VWM is especially suitable for treatment with iPSCs as this allows genetic modification of the mutation. However, the current procedure for developing iPSCs lacks safety and efficiency in order to get enough transplantable cells.

No animal studies or clinical trials for cell replacement therapy in VWM have been done to date. Before starting clinical trials, proof-of-principle studies in experimental animal models are needed. A mouse model for VWM has been reported by Geva et al. (2010). Analysis of the mice revealed impaired motor functions and altered growth rate, but under normal conditions the mice did not develop severe clinical symptoms and had a normal life span. Currently, mouse models for VWM are being developed and characterized by the Van der Knaap & Heine labs, carrying mutations associated with a more severe disease course than the mouse model of Geva et al. (2010). Preliminary results show that *eif2b* mutants develop ataxia and epilepsy which resemble the clinical signs of VWM patients. Histopathological analysis indicates abnormal white matter structures and presence of dysmorphic astrocytes in adult mice. Transplantation studies are underway to investigate whether cell grafts can alleviate this VWM pathology.

Graft studies in mouse models are expected to give insight into several matters. Firstly, in what stage of the disease should cell therapy take place. Transplantation in an early stage might prevent further damage, but the risk of damaging the brain must be taken into account. Furthermore, in an early stage glial cells of the patient may still be functional in large areas and prevent grafting of the transplanted cells in these areas, which might become affected in later stages. In this case, additional treatment will be required. Multiple injections into the brain should be avoided as much as possible. On the other hand, if the cell therapy is given too late, the white matter structure may be damaged too much to allow new glial cells to integrate.

Initial grafting studies should target replacement of oligodendrocyte, astrocyte or common glia precursor cells, as these are mainly affected in VWM patients. As discussed before, the Van der Knaap lab found that the astrocyte population looks dysmorphic and is thought to disturb oligodendrocyte maturation and therefore proper myelination. So although mutated oligodendrocytes are unable to myelinate white matter areas, replacement of the astrocytic cell population could be the best strategy. However more research is necessary to find out which cell population is most suitable for treatment: astrocyte precursor cells (APCs), APCs plus OPCs or glial precursor cells (GPCs). Animal models will further resolve the question whether the VWM brain is susceptible to cell grafts and if other cell environmental factors will play a role in successful integration of the grafted cells. In vitro co-culture systems with *eif2b* mutant cells will likely help resolving these questions.

3.4 Conclusion

Proof-of-principle studies in experimental animals for PD have shown that cell transplants of SC-derived or fetal neural cells can improve neurological function, and therefore have potential for human treatment. Although current medications for PD initially work for most patients, cell replacement therapy could help those patients that do not benefit from L-dopa anymore or suffer from severe side effects. Improved protocols are needed to improve efficiency of cell production, to generate more pure populations of transplantable cells and to increase survival of cell grafts. Another issue involves the usability of iPSCs in cell therapies for PD. Although there are familial forms of PD known, most cases are sporadic. So iPSCs derived from the PD patient, having the same genetic risk factors and undergone the same environmental influences, may develop PD pathology faster than transplanted cells derived from healthy donors. Comparative animal studies will hopefully resolve the question whether ESCs or iPSCs will be more suitable for cell therapy in PD patients, taking into account the risk of eliciting an immune response with ESCs compared with the risk of developing PD pathology in iPSCs.

New mouse models for VWM will help answering many questions about the onset and possibility to develop cell replacement therapies. VWM patients have very poor prospects. So if cell grafts could only prolong life expectancy or improve VWM pathology associated with a risk of developing cancer, transplantation could become an important new treatment option for these patients.

In conclusion, different neurodegenerative diseases require different hurdles to be overcome before cell replacement therapy can become a realistic clinical strategy. In all cases SC research in both basic and preclinical settings is still necessary.

References

Albin RL, Young AB, Penney JB (1995) The functional anatomy of disorders of the basal ganglia. Trends Neurosci 18:63–64

Bankiewicz KS, Plunkett RJ, Jacobowitz DM, Porrino L, di PU, London WT, Kopin IJ, Oldfield EH (1990) The effect of fetal mesencephalon implants on primate MPTP-induced parkinsonism. Histochemical and behavioral studies. J Neurosurg 72:231–244

Braak H, Braak E (2000) Pathoanatomy of Parkinson's disease. J Neurol 247(Suppl2):II3–10

Brundin P, Nilsson OG, Strecker RE, Lindvall O, Astedt B, Bjorklund A (1986) Behavioural effects of human fetal dopamine neurons grafted in a rat model of Parkinson's disease. Exp Brain Res 65:235–240

Bugiani M, Boor I, van Kollenburg B, Postma N, Polder E, van Berkel C, van Kesteren RE, Windrem MS, Hol EM, Scheper GC, Goldman SA, van der Knaap MS (2011) Defective glial maturation in vanishing white matter disease. J Neuropathol Exp Neurol 70:69–82

Chaudhuri KR, Healy DG, Schapira AH (2006) Non-motor symptoms of Parkinson's disease: diagnosis and management. Lancet Neurol 5:235–245

Cooper O, Hargus G, Deleidi M, Blak A, Osborn T, Marlow E, Lee K, Levy A, Perez-Torres E, Yow A, Isacson O (2010) Differentiation of human ES and Parkinson's disease iPS cells into ventral midbrain dopaminergic neurons requires a high activity form of SHH, FGF8a and specific regionalization by retinoic acid. Mol Cell Neurosci 45:258–266

de Rijk MC, Tzourio C, Breteler MM, Dartigues JF, Amaducci L, Lopez-Pousa S, Manubens-Bertran JM, Alperovitch A, Rocca WA (1997) Prevalence of parkinsonism and Parkinson's disease in Europe: the EUROPARKINSON collaborative study. European Community concerted action on the epidemiology of Parkinson's disease. J Neurol Neurosurg Psychiatry 62:10–15

Eichler F, Van Haren K (2007) Immune response in leukodystrophies. Pediatr Neurol 37:235–244

Freed CR (2002) Will embryonic stem cells be a useful source of dopamine neurons for transplant into patients with Parkinson's disease? Proc Natl Acad Sci USA 99:1755–1757

Gelb DJ, Oliver E, Gilman S (1999) Diagnostic criteria for Parkinson disease. Arch Neurol 56:33–39

Geva M, Cabilly Y, Assaf Y, Mindroul N, Marom L, Raini G, Pinchasi D, Elroy-Stein O (2010) A mouse model for eukaryotic translation initiation factor 2B-leucodystrophy reveals abnormal development of brain white matter. Brain 133:2448–2461

Gibb WR, Lees AJ (1989) The significance of the Lewy body in the diagnosis of idiopathic Parkinson's disease. Neuropathol Appl Neurobiol 15:27–44

Hargus G, Cooper O, Deleidi M, Levy A, Lee K, Marlow E, Yow A, Soldner F, Hockemeyer D, Hallett PJ, Osborn T, Jaenisch R, Isacson O (2010) Differentiated Parkinson patient-derived induced pluripotent stem cells grow in the adult rodent brain and reduce motor asymmetry in Parkinsonian rats. Proc Natl Acad Sci USA 107:15921–15926

Hedlund E, Perlmann T (2009) Neuronal cell replacement in Parkinson's disease. J Intern Med 266:358–371

Hwang DY, Kim DS, Kim DW (2010) Human ES and iPS cells as cell sources for the treatment of Parkinson's disease: current state and problems. J Cell Biochem 109:292–301

Kawasaki H, Suemori H, Mizuseki K, Watanabe K, Urano F, Ichinose H, Haruta M, Takahashi M, Yoshikawa K, Nishikawa S, Nakatsuji N, Sasai Y (2002) Generation of dopaminergic neurons and pigmented epithelia from primate ES cells by stromal cell-derived inducing activity. Proc Natl Acad Sci USA 99:1580–1585

Kim SU, de Vellis VJ (2009) Stem cell-based cell therapy in neurological diseases: a review. J Neurosci Res 87:2183–2200

Koch P, Kokaia Z, Lindvall O, Brustle O (2009) Emerging concepts in neural stem cell research: autologous repair and cell-based disease modelling. Lancet Neurol 8:819–829

Kordower JH, Freeman TB, Snow BJ, Vingerhoets FJ, Mufson EJ, Sanberg PR, Hauser RA, Smith DA, Nauert GM, Perl DP (1995) Neuropathological evidence of graft survival and striatal reinnervation after the transplantation of fetal mesencephalic tissue in a patient with Parkinson's disease. N Engl J Med 332:1118–1124

Labauge P, Horzinski L, Ayrignac X, Blanc P, Vukusic S, Rodriguez D, Mauguiere F, Peter L, Goizet C, Bouhour F, Denier C, Confavreux C, Obadia M, Blanc F, de SJ, Fogli A, Boespflug-Tanguy O (2009) Natural history of adult-onset eIF2B-related disorders: a multi-centric survey of 16 cases. Brain 132:2161–2169

Lee SH, Lumelsky N, Studer L, Auerbach JM, McKay RD (2000) Efficient generation of midbrain and hindbrain neurons from mouse embryonic stem cells. Nat Biotechnol 18:675–679

Lindvall O, Widner H, Rehncrona S, Brundin P, Odin P, Gustavii B, Frackowiak R, Leenders KL, Sawle G, Rothwell JC (1992) Transplantation of fetal dopamine neurons in Parkinson's disease: one-year clinical and neurophysiological observations in two patients with putaminal implants. Ann Neurol 31:155–165

Luquin MR, Scipioni O, Vaamonde J, Gershanik O, Obeso JA (1992) Levodopa-induced dyskinesias in Parkinson's disease: clinical and pharmacological classification. Mov Disord 7:117–124

Maries E, Dass B, Collier TJ, Kordower JH, Steece-Collier K (2003) The role of alpha-synuclein in Parkinson's disease: insights from animal models. Nat Rev Neurosci 4:727–738

Montastruc JL, Llau-Bousquet ME, Senard JM, Rascol O (1997) Movement disorders of drug origin. Rev Prat 47:1109–1116

Morizane A, Li JY, Brundin P (2008) From bench to bed: the potential of stem cells for the treatment of Parkinson's disease. Cell Tissue Res 331:323–336

Olanow CW, Goetz CG, Kordower JH, Stoessl AJ, Sossi V, Brin MF, Shannon KM, Nauert GM, Perl DP, Godbold J, Freeman TB (2003) A double-blind controlled trial of bilateral fetal nigral transplantation in Parkinson's disease. Ann Neurol 54:403–414

Papadeas ST, Maragakis NJ (2009) Advances in stem cell research for Amyotrophic Lateral Sclerosis. Curr Opin Biotechnol 20:545–551

Park CH, Minn YK, Lee JY, Choi DH, Chang MY, Shim JW, Ko JY, Koh HC, Kang MJ, Kang JS, Rhie DJ, Lee YS, Son H, Moon SY, Kim KS, Lee SH (2005) In vitro and in vivo analyses of human embryonic stem cell-derived dopamine neurons. J Neurochem 92:1265–1276

Preynat-Seauve O, Burkhard PR, Villard J, Zingg W, Ginovart N, Feki A, Dubois-Dauphin M, Hurst SA, Mauron A, Jaconi M, Krause KH (2009) Pluripotent stem cells as new drugs? The example of Parkinson's disease. Int J Pharm 381:113–121

Rhee YH, Ko JY, Chang MY, Yi SH, Kim D, Kim CH, Shim JW, Jo AY, Kim BW, Lee H, Lee SH, Suh W, Park CH, Koh HC, Lee YS, Lanza R, Kim KS, Lee SH (2011) Protein-based human iPS cells efficiently generate functional dopamine neurons and can treat a rat model of Parkinson disease. J Clin Invest 121:2326–2335

Soldner F, Hockemeyer D, Beard C, Gao Q, Bell GW, Cook EG, Hargus G, Blak A, Cooper O, Mitalipova M, Isacson O, Jaenisch R (2009) Parkinson's disease patient-derived induced pluripotent stem cells free of viral reprogramming factors. Cell 136:964–977

Spillantini MG, Schmidt ML, Lee VM, Trojanowski JQ, Jakes R, Goedert M (1997) Alpha-synuclein in Lewy bodies. Nature 388:839–840

Takagi Y, Takahashi J, Saiki H, Morizane A, Hayashi T, Kishi Y, Fukuda H, Okamoto Y, Koyanagi M, Ideguchi M, Hayashi H, Imazato T, Kawasaki H, Suemori H, Omachi S, Iida H, Itoh N, Nakatsuji N, Sasai Y, Hashimoto N (2005) Dopaminergic neurons generated from monkey embryonic stem cells function in a Parkinson primate model. J Clin Invest 115:102–109

van der Knaap MS, Pronk JC, Scheper GC (2006) Vanishing white matter disease. Lancet Neurol 5:413–423

van der Knaap MS, van Berkel CG, Herms J, Van CR, Baethmann M, Naidu S, Boltshauser E, Willemsen MA, Plecko B, Hoffmann GF, Proud CG, Scheper GC, Pronk JC (2003) eIF2B-related disorders: antenatal onset and involvement of multiple organs. Am J Hum Genet 73:1199–1207

Wernig M, Zhao JP, Pruszak J, Hedlund E, Fu D, Soldner F, Broccoli V, Constantine-Paton M, Isacson O, Jaenisch R (2008) Neurons derived from reprogrammed fibroblasts functionally integrate into the fetal brain and improve symptoms of rats with Parkinson's disease. Proc Natl Acad Sci USA 105:5856–5861

Wijesekera LC, Leigh PN (2009) Amyotrophic lateral sclerosis. Orphanet J Rare Dis 4:3

Wijeyekoon R, Barker RA (2009) Cell replacement therapy for Parkinson's disease. Biochim Biophys Acta 1792:688–702

Winkler C, Kirik D, Bjorklund A (2005) Cell transplantation in Parkinson's disease: how can we make it work? Trends Neurosci 28:86–92

Witt J, Marks WJ Jr (2011) An update on gene therapy in Parkinson's disease. Curr Neurol Neurosci Rep 11:362–370

Yasuhara T, Matsukawa N, Hara K, Yu G, Xu L, Maki M, Kim SU, Borlongan CV (2006) Transplantation of human neural stem cells exerts neuroprotection in a rat model of Parkinson's disease. J Neurosci 26:12497–12511

Yoshizaki T, Inaji M, Kouike H, Shimazaki T, Sawamoto K, Ando K, Date I, Kobayashi K, Suhara T, Uchiyama Y, Okano H (2004) Isolation and transplantation of dopaminergic neurons generated from mouse embryonic stem cells. Neurosci Lett 363:33–37

Chapter 4
Conclusions

Abstract Recent developments in SC research have driven a rush to find the best iPSC protocol, increasing studies into reprogramming efficiency and animal transplantation. However, limits to the clinical use of iPSCs remain. It is also important to continue parallel studies into ESCs, given its longer history and consequently more advanced knowledge base. Both fields of research stand to benefit from each other, and so personalized SC therapies will develop more rapidly using a combined approach.

Keywords iPSC · ESC · Trans-differentiation · Regenerative medicine · Disease modelling · Reprogramming method · Epigenetics

4.1 The Promise and Limitations of iPSCs

Disease and patient-specific iPSCs provide a tremendous opportunity for customized medical investigation and therapy development. The arrival of iPSCs eliminated some important dilemmas concerning the use of ESCs in regenerative medicine. First, the ethical controversy regarding use of human embryos, which is necessary to obtain ESCs from the ICM of blastocysts, would no longer exist. Second, iPSCs offer a practical advantage over ESCs, by generating patient-specific pluripotent SCs, which could overcome concerns about immune rejection. Recent advances in gene therapy would also allow iPSC treatment for genetically determined diseases. Generation of disease-specific iPSCs will provide *human* disease models. Animal models are very useful, but findings from these models do not always apply to human pathophysiology. Although only in vitro, experiments may be conducted on disease phenotypes, including potential drug treatments, thereby reducing or removing the safety issues inherent to clinical trials.

However, there are still limitations to using iPSCs in clinical settings. The first is associated with the production of iPSCs in a safe and efficient manner. Viral transduction of the OSKM TFs is efficient but the risk of genome integration is high. Genomic integration can disrupt gene function and cause tumour formation. Furthermore, re-expression of c-Myc and, to a lesser extent, Klf4 upon differentiation is associated with tumour formation as well. Therefore, it is important that methods are developed to produce iPSCs free of genomic integration. Given recent successes using non-viral approaches, it is probably only a matter of time before safe reprogramming protocols are available. For example, studies have already reported iPSC reprogramming using miRNA's and recombinant cell-penetrating proteins (Miyoshi et al. 2011; Zhou et al. 2009).

Feeder cells are often used to induce neural fate, and these cells are often non-human (Morizane et al. 2008), which is a criterion for exclusion for clinical trials in patients. Animal products and cells are a potential source of disease-causing pathogens. During the process of iPSC derivation animal products are often used, but Rodriguez-Piza et al. (2010) showed that they can produce iPSCs without the use of animal-derived products. However, the precise mechanism by which the neural induction by feeder cells is accomplished is not well understood, which makes optimization of feeder cells and co-culture protocols very complicated. A thorough investigation of the exact proteins and substances that are important for differentiation is necessary in order to culture neural cells without the use of animal derived products or cells. The development of safe and robust differentiation and selection methods, however, will likely need more time.

So far, ESCs have been used to evaluate the epigenetic status of iPSCs. Several epigenetic differences have been described, but it is not known what the effect of these differences will be on the differentiation or tumourigenicity of the cells (Barrilleaux and Knoepfler 2011a). Furthermore, differences between ESCs and iPSCs seem to diminish with prolonged culturing of iPSCs. More research is necessary to identify the amount of epigenetic memory still present in iPSCs and to investigate the effect of the epigenetic state on differentiation and the functioning of iPSC-derived cells. It may be easier for an iPSC to differentiate into a cell type of the same lineage of the source cell than into a completely different cell type. Some reprogramming methods are more efficient in erasing the epigenetic memory than others. It will be necessary to choose a reprogramming method not only for the reprogramming efficiency but also for the ability to erase the epigenetic memory of the cells.

More specific to regenerative medicine, assays such as chimera formation and contribution to germ line are available to qualify full reprogramming for other species but not for human tissue. Molecular based assays, focused on the epigenomes of cells, have been shown to offer alternatives that could be used for humans. They are also faster, and increasingly more accurate to defining cell identity than functional and gene expression data. This is particularly true now that gene states are revealed to be subtler than "off" and "on", with markings that cannot be determined by expression levels alone. As genome wide chromatin mapping becomes broader, including different cell types, ages, and states of health, the minimal conditions for safe, competent tissue for regenerative therapies should emerge. Coinciding with this line of research, the discovery and/or development of epigenetic control mechanisms, especially active mechanisms within oocytes and embryos, hold the potential to remedy current efficiency and safety concerns.

A study by Zhao et al. (2011a) showed that transplantation of iPSCs did result in an immune response. This was quite disturbing news since iPSCs is meant to overcome the risk of rejection which is related to ESCs. The question is why these iPSCs would trigger an immune response, given that they are genetically similar to the host. Reprogramming with episomal vectors led to a similar immune response, so it is not related to re-expression of OSKM or transgene integration. The authors found an overexpression of a small number of genes in the iPSC-derived teratomas (Zhao et al. 2011a). Three of these genes also led to an immune response in matched ESCs, which do not normally activate the immune system. It is not known why these genes become activated or whether these genes would still be a problem when transplanting differentiated cells derived from iPSCs. A recent review about iPSCs argued that the activation of these genes might be related to tumour formation (Barrilleaux and Knoepfler 2011b). Certain tumours are highly immunogenic, and at least one of the genes (*Hormad1*) found by Zhao et al. 2011b is known to be a tumour-specific antigen. Notably, previous studies have shown successful transplantation of iPSC derived cells in mice without any rejection issues (Hanna et al. 2007). However, before therapies are developed, it is necessary to elucidate if differentiated iPSCs have the potential to trigger an immune response as well.

We should also be aware of novel ethical issues arising from this alternative approach. For example, patient-specific iPSCs increase the possibility of using derived cells without consent of the donor, including production of gametes for reproductive purposes. Therefore, strict rules should be made, such as limiting production of gametes to in vitro studies, and requiring signed patient consent forms which clearly define acceptable uses for collected and derived cell lines.

Another concern involves the high cost associated with such therapies, which most health care systems and many patients may not be able to afford. A standard set of iPSC or ESC lines could provide close immunological matches for large groups of people (Taylor et al. 2011). However, the feasibility of such a strategy still needs to be assessed. But if this holds true, the reprogramming field should increase focus on generating disease-specific iPSCs that offer insights into neurodegenerative and pathological processes, to facilitate drug development.

Another limitation concerns the application of iPSCs for sporadic diseases like the majority of PD cases. In such diseases, there are often genetic and environmental risk factors that jointly cause the disease pathology. These risk factors are often unknown, or have only a moderate effect. Somatic cells derived from the patient would contain identical genetic risk factors as the affected cells and environmental risk factors may also remain the same. For example, it is possible that there are brain-specific environmental risk factors, or that the disease pathology is associated with a long-term accumulation of proteins that are not present in the somatic cells. In these cases, it may take a number of years before cell transplants become pathological, and they may even help patients before disease development. Ultimately, sporadic diseases are more likely to effect patient specific replacement cells than non-autologous tissue, and so form a reason to use ESCs instead of iPSCs.

4.2 Concluding Remarks

The first iPSCs were developed just five years ago. Since then, this line of research has evolved very fast, having shown great promise for use in regenerative medicine. It appears likely that as cell reprogramming techniques progress, production of iPSCs and direct trans-differentiation will play complementary roles. While trans-differentiation is capable of direct and in vivo conversion to desired cell types, iPSCs are arguably more useful when tissues must be re-engineered to remove genetic defects, large quantities are desired, or for disease-specific developmental modelling. Both methods currently suffer from limited efficiency, safety risks, and lack of clear criteria for determining the success of cell reprogramming. However it can be expected that many of the current limitations will be dealt with in the future, just as previous issues have been addressed. In moving forward with iPSCs it is important that ESCs, and the vast experience accrued from them, will not be forgotten. Near and long term goals related to making regenerative medicine a practical reality require knowledge from active ESC and reprogramming research.

References

Barrilleaux B, Knoepfler PS (2011a) Inducing iPSCs to escape the dish. Cell Stem Cell 9: 103–111

Barrilleaux B, Knoepfler PS (2011b) Inducing iPSCs to escape the dish. Cell Stem Cell 9: 103–111

Hanna J, Wernig M, Markoulaki S, Sun CW, Meissner A, Cassady JP, Beard C, Brambrink T, Wu LC, Townes TM, Jaenisch R (2007) Treatment of sickle cell anemia mouse model with iPS cells generated from autologous skin. Science 318:1920–1923

Miyoshi N, Ishii H, Nagano H, Haraguchi N, Dewi DL, Kano Y, Ni-shikawa S, Tanemura M, Mimori K, Tanaka F, Saito T, Ni-shimura J, Takemasa I, Mizushima T, Ikeda M, Yamamoto H, Sekimoto M, Doki Y, Mori M (2011) Reprogramming of mouse and human cells to pluripotency using mature mi-croRNAs. Cell Stem Cell 8:633–638

Morizane A, Li JY, Brundin P (2008) From bench to bed: the poten-tial of stem cells for the treatment of Parkinson's disease. Cell Tissue Res. 331:323–336

Rodriguez-Piza I, Richaud-Patin Y, Vassena R, Gonzalez F, Barrero MJ, Veiga A, Raya A, Belmonte JC (2010) Reprogramming of human fibroblasts to induced pluripotent stem cells under xeno-free conditions. Stem Cells 28:36–44

Taylor CJ, Bolton EM, Bradley JA (2011) Immunological considerations for embryonic and induced pluripotent stem cell banking. Philos Trans R Soc Lond B Biol Sci 366:2312–2322

Zhao T, Zhang ZN, Rong Z, Xu Y (2011a) Immunogenicity of in-duced pluripotent stem cells. Nature 474:212–215

Zhao T, Zhang ZN, Rong Z, Xu Y (2011b) Immunogenicity of in-duced pluripotent stem cells. Nature 474:212–215

Zhou H, Wu S, Joo JY, Zhu S, Han DW, Lin T, Trauger S, Bien G, Yao S, Zhu Y, Siuzdak G, Scholer HR, Duan L, Ding S (2009) Generation of induced pluripotent stem cells using recombinant proteins. Cell Stem Cell 4:381–384